99 Ways To Improve Your HI-FI

by
Len Buckwalter

HOWARD W. SAMS & CO., INC.
THE BOBBS-MERRILL CO., INC.
INDIANAPOLIS · KANSAS CITY · NEW YORK

FIRST EDITION

FOURTH PRINTING—1974

Copyright © 1971 by Howard W. Sams & Co., Inc., Indianapolis, Indiana 46268. Printed in the United States of America.

All rights reserved. Reproduction or use, without express permission, of editorial or pictorial content, in any manner, is prohibited. No patent liability is assumed with respect to the use of the information contained herein. While every precaution has been taken in the preparation of this book, the publisher assumes no responsibility for errors or omissions. Neither is any liability assumed for damages resulting from the use of the information contained herein.

International Standard Book Number: 0-672-20876-8
Library of Congress Catalog Card Number: 78-175569

Preface

The day a brand-new hi-fi system is brought home, it is ready for improvement. Even the manufacturer who produces state-of-the-art equipment cannot carry all his skills beyond the factory. Equipment must be installed, operated, and heard under a myriad of unpredictable conditions in millions of homes. A speaker, for example, may perform superbly in the anechoic chamber of a laboratory, yet sound tinny and hollow in a living room. An amplifier may deliver flawless power in the quality-control department, but distort after operating for a few months in the overheated corner of a living-room shelf. All the precision built into a phonograph cartridge may be destroyed by a tiny dust blob.

These situations are three reasons for this book. The 96 others are similarly intended to keep a hi-fi healthy, to help it deliver all its inherent quality, and to demonstrate dozens of techniques that cannot be jammed into the instruction manual. Some items, like those on extension speakers, show how to derive additional enjoyment from a substantial hi-fi investment. If hundreds of dollars are spent for a superb sound in the living room, why not bring its pleasure into other areas, even outdoors, at a tiny fraction of the original cost.

Most such improvements can be done by the nontechnical person without special skill or tools. Several items slanted at the hobbyist use some simple test equipment. Whatever your proficiency, it is a good idea to have the manufacturer's literature for your specific equipment. It contains the last word on such matters as lubrication and certain other practices that can be described here only in a general way. A manufacturer might have some special precaution; some solid-state stereo amplifiers, for example, have a restriction on connecting together speaker grounds. A warning to this effect should appear prominently in the owner's instruction.

But in the majority of installations, the tips, hints, and techniques described on these pages should prove helpful to anyone seeking more enjoyment from their hi-fi.

LEN BUCKWALTER

Contents

SECTION 1. Installation

1. Better Bass in Corner 9
2. Avoid Mechanical Feedback 9
3. Standing Waves 10
4. Speaker Phasing 12
5. Cartridge Hum 13
6. Speaker Location 14
7. Front-to-Back Sound 15

SECTION 2. FM Stereo Reception

8. Multipath Distortion 17
9. FM From TV Antenna 20
10. Constructing an FM Dipole 21
11. Better Twin Lead 22
12. Coaxial Cable 23
13. Preamplifier 25
14. Front-End Overload 26
15. FM Bandpass Filter 26
16. Antenna Connection 27
17. Tuning Tips 27

SECTION 3. Phonos and Turntables

18. Cleaning Phonograph Records 29
19. Speed Accuracy 30
20. Deburr the Disc 32

21. Stylus Force 32
22. Stylus Dust 33
23. Keep Phono Drive Clean 34
24. Cartridge Mount 36
25. Turntable Lube 37
26. Loose Belts 38
27. Master Shut-Off 39
28. Leveling the Turntable Base 40

SECTION 4. Amplifiers

29. Quiet Location 43
30. VU Meter 43
31. Dummy Load 45
32. Voltmeter Checks Watts 46
33. Frequency Response 48
34. Electrolytics 50

SECTION 5. Tape Equipment

35. Less Print-Through 53
36. Nice Splice 54
37. Broken Tape 56
38. Clean Heads 57
39. Degaussing 57
40. Mechanical Noises 58
41. Pressure Pads 59
42. Replacing Heads 60
43. Tape Speed 62
44. Head Alignment 63
45. Excessive Hum 64
46. Wow and Flutter 64
47. Rubber Belts 65

SECTION 6. Loudspeakers

48. Mounting Board 67
49. Speaker Wires 68
50. Extension Speakers (Parallel) 69
51. Extension Speakers (Series) 69
52. Long-Distance Speaker 70

53. Main-Remote Selector	71
54. Remote Volume Control	72
55. Remote Stereo Balance Control	74
56. Speaker Switching	74
57. Impedance Selector	75
58. Torn Paper Cone	77
59. Gritty Voice Coil	78
60. Correcting a Cheap Enclosure	79

SECTION 7. Headphones and Mikes

61. Choosing Headphones	81
62. Add a Headphone Jack	82
63. Home-Made Stereo Headphones	83
64. Long Mike Cables	87
65. Choosing a Mike Pattern	88
66. Extra Mike	89
67. Mike Noises	90

SECTION 8. Techniques

68. Free White Noise	93
69. Reversing AC Plugs	94
70. Cable Coding	94
71. Assembling Phono Plugs	95
72. Trade-In	97
73. Give It More Air	98
74. Switching Surges	99
75. Shifting Line Voltage	100
76. Accidental Tape Erasure	101
77. Sticky Cassette	102

SECTION 9. Accessories

78. Simple Crossover	103
79. Test Record	104
80. Hiss Reduction	105
81. Audio Mixer	106
82. A-M Radio Adapter	107
83. Cooling Fan	109

84. Recording Adapters 110
85. Station Guide 111

SECTION 10. Troubleshooting Tips

86. Intermittent Kits 113
87. Tube Testing 113
88. Cartridge Terminals 115
89. Ground Loops 115
90. Hidden Cable Breaks 116
91. Scratchy Controls 117
92. Free Test Set 118

SECTION 11. Interference

93. Electrical Ground 119
94. CB-Ham Voices 120
95. Transmitter Filter 122
96. Arcing Motors 123
97. Appliance Bypasses 123
98. Aircraft Interference 124
99. High-Pass Filter 127

SECTION 1

Installation

1
Better Bass in Corner

The base response of a loudspeaker can be improved by placing the speaker enclosure in the corner of a room where the walls meet. The walls extend outward and form the mouth of a huge horn, with the speaker acting as the driving source. The effect raises the acoustic load on the speaker and couples more audio energy into the surrounding air. This arrangement is especially useful for reinforcing the bass tones of a small speaker.

Although the best corner to locate a speaker is one where the walls meet, do not overlook other areas that might create the horn loading effect—where a wall and ceiling meet, or at the junction of a partition or room divider and the floor. If the speaker cabinet is portable, a convincing experiment is to move it toward a corner while it is playing. The improvement noticed in the last few feet can be surprising.

2
Avoid Mechanical Feedback

A troublespot to guard against in any installation is a solid path between the loudspeaker and turntable. The effects can be anything from booming thunder as you walk across the room, to muddying of music played on discs. The distortion usually grows worse as the bass or volume control is advanced, or when program material is rich in low tones.

Two assets of a hi-fi system—high amplification and good low-frequency response—are at the root of the problem. A random distubance in the room might reach the phono needle and vibrate the pickup. The noise is amplified, then emitted by the loudspeaker as a growl or rumble. The speaker cabinet physically shakes and sends a vibration back to the needle. The process is thereby repeated—the needle shakes, the noise is amplified, and so on. The seriousness of the problem depends on how easily speaker vibrations can travel over the feedback path to the needle.

In extreme cases, a continuous oscillation builds quickly and damages the speakers. More often, there is a rumble that dies away after the triggering source (usually footsteps) is removed. If these obvious symptoms are not present, you might be able to detect feedback that is introducing a subtle deterioration in the music. Stop the turntable from rotating, but keep the needle on the disc. Amplifier volume and tone controls should be turned to the highest position you use. Listen critically for any sound from the speaker while walking in the room. The speaker should not echo your footfalls. Rap your knuckles on the shelf or cabinet that holds the phonograph. The pickup should react with little or no sensitivity to the mechanical disturbance. If there is an annoying sound, take the following steps to kill the solid feedback path from loudspeaker to phono pickup.

Never mount the phonograph directly on a speaker cabinet. This almost always guarantees a solid loop through which vibrations feed back into the system.

Placing the speaker and phono on the same shelf may also create a favorable path for feedback. Use separate shelves if feasible.

A path can also occur between a speaker and phono that are separated, but in contact with the floor through their cabinets. Mounting the phono only on a wall shelf may solve this one.

Finally, a foam pad placed under the phono base and/or speaker cabinets can interrupt the feedback path. Some experimentation will show which measures you will need to take.

3

Standing Waves

Sound waves emanating from a loudspeaker can bounce from a wall and produce reflected energy which bucks out part of the original wave. In the listening room it causes areas of both weakened and reinforced sound. The effect, most pronounced toward the lower end of the audio spectrum, is caused by standing waves. How they are formed is shown in Fig. 1-1. For the sake of illustration, the speaker is generating a pure 400-Hz tone. This creates a wave approximately 33 inches long, which is the physical distance between two air compressions in the sound wave. As the wave travels and strikes the wall, returning

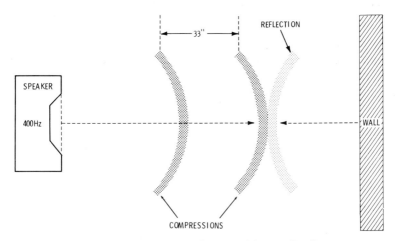

Fig. 1-1. A standing wave is caused by a reflection.

energy encounters the original wave. If the two waves are out of phase where they meet, the compression of one cancels the rarefaction of the other. It creates a "null" area of no sound. If the speaker or the wall were moved by 16½ inches in either direction, the reflected wave would produce the opposite effect, and the two compressions would add and reinforce.

In an actual room, no such pure situation exists. Audio wavelengths are constantly changing with frequency, as well as distances needed to produce these effects. Nevertheless, standing waves are considered an important factor in the small listening room.

What to do about it? In our example, it took only a distance of 16½ inches to change from cancellation to reinforcement. The dimension is greater for lower tones, smaller for higher ones. In a practical situation, however, poor sound caused by standing waves can be helped by experimentally relocating the speakers. You will discover that even small differences in the position of the speaker can improve the weak low tones. Make small changes, then walk about the listening area to determine the result. (Be sure to also check the item about speaker phasing; it also can produce a similar effect.)

4
Speaker Phasing

When speakers are mounted close to each other the waves of air pressure emanating from their cones mix. The action may create dead spots in the listening room, loss of certain tones, and poor stereo separation. Speakers operate out of phase when the cone of one moves in, while the other pushes outward, causing cancellation of air pressure and loss of sound. There are several ways to ensure that speakers are operating in unison; that is, in phase.

Assume your speakers are facing in the same general direction. This will be the situation in most home stereo installations. If you have some unorthodox arrangement, where cones face each other, an in-phase hookup might make matters worse. For these cases use the steady-tone method, described later, to judge the best phasing connection. But, in general, most speakers point in the direction of the listener and call for the in-phase treatment.

Some manufacturers mark speaker terminals with a + symbol or color dot as a guide to phasing. If you keep all speakers wired in identical fashion—the same amplifier taps going to the same speaker terminals—chances are phasing will be correct. It's always a good idea to check phasing anyway. There could be some element (i.e. a crossover network) that changes audio phase in the system. Phasing is easy to do and the effort repays itself in better speaker operation.

If you have a critical ear, you may hear the difference in speaker phasing. Since low tones suffer most from incorrect phasing, play music (monophonically) that is rich in low frequencies. After listening a while, reverse the leads to *one* speaker. This can be done at the speaker or amplifier terminals, but not both. Listen again to see if the bass is reinforced. In some amplifiers, a phasing switch lets you do the same thing without manually reconnecting wires.

If you have a test record or other source of a low-frequency steady tone (like an audio oscillator) it is easier to hear if speaker phasing is correct. With the tone playing equally in both speakers, change the connection to one speaker and determine best bass. Another test is to position yourself between the two speakers and stand several feet away. With equal tones in the speakers, the apparent source should be a central point between them. This is important for good separation when playing stereo program material.

If you have unmounted speakers, or speakers in a cabinet with a removable back, you can use an old trick with an ordinary 1½-volt flashlight cell (any size) to determine phasing. Touch the positive, or button, side of the battery to one speaker terminal. Connect the flat, or negative, side of the battery to the other speaker terminal. As you make contact, the speaker cone will move with a click or thump. You want it to move in— toward the back of the speaker. If it does, mark one speaker terminal with a + symbol, as shown in Fig. 1-2.

If the hookup does not move the cone inward, reverse the battery connection. Now mark the + terminal on the speaker. With this procedure you can use the battery to mark all your speakers and connect their terminals to the stereo amplifier

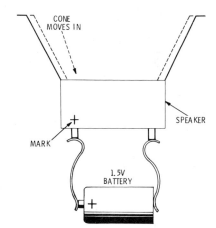

Fig. 1-2. Phasing a speaker.

in comparable fashion. In small speakers you may not see cone movement; the action is too limited. If this happens, gently place your fingertips on the cone to feel the direction of movement. This technique, in fact, might be the only one to determine the direction of cone movement in a speaker that is mounted inside a cabinet.

If you make any future changes in your system, like installing a new phono cartridge, preamplifier, or crossover network, recheck speaker phasing. It might be inadvertently reversed.

5

Cartridge Hum

The small white object seen held in the hand in Fig. 1-3 is a typical magnetic cartridge, or phono pickup. It has fine turns of wire which form internal coils. The large dark object just behind it is a power transformer, another device containing coils. Both devices are usually shielded, but there can be hum pickup between a cartridge and transformer. A power transformer carries heavy currents which radiate electromagnetic hum fields.

Because of a coupling possibility, always keep a phono tone arm as far as conveniently possible from an amplifier power transformer. Mere inches may suffice in some cases, but in others several feet may be necessary. A way to check for hum

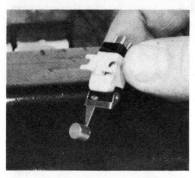

Fig. 1-3. Typical magnetic cartridge.

pickup is to set the amplifier volume to normal listening level and the function switch to magnetic phono input. If hum changes with cartridge position, additional spacing may provide the cure. Also, check if the phonograph is properly grounded as described elsewhere in this section. If grasping a tone arm or cartridge shell changes hum level, improved grounding might reduce it.

6

Speaker Location

Experiment with speaker position since small shifts in room location may produce large changes in sound quality. Room size, furnishings, and listening position are all at work. Carpets, stuffed chairs, drapes, and other absorptive surfaces deaden sound, while bare floors and hard areas give it a bright, live quality. (The effects of standing waves and corner locations are described elsewhere in this section.)

A starting point for speaker separation is a spacing of about 6 to 8 feet. Sit in the listening chair and have someone vary the distance between speakers until you experience the most pleasing stereo effect. If the room is small, or tonal range somehow deficient, adjust the amplifier tone, volume, or balance controls to help offset room acoustics.

In speaker cabinets the tweeter might be mounted off center. If high tones are poor, you may have placed the tweeter nearest to the floor, and the treble tones are becoming lost in some obstruction. It is usually best to place a tweeter in the topmost position. Also, alternately angle the speakers slightly toward and away from each other to check for best performance.

7
Front-to-Back Sound

This simple setup, devised by David Hafler, a noted authority on hi-fi, captures some of the "hall sound" experienced at a live concert. Operating with a standard two-channel stereo amplifier, it reconstructs certain information from left and right

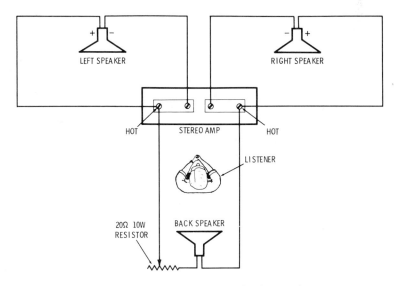

Fig. 1-4. Hookup for front-to-back sound.

program channels and feeds it to a third speaker positioned to the rear of the listener (Fig. 1-4). Although the effect varies according to how the particular recording was made, the third speaker imparts a feeling of spaciousness and reverberation not recovered by conventional front stereo speakers.

One precaution in this hookup should be observed when using speakers of less than 8 ohms. Some transistor amplifiers may not be safely connected to speakers of lower impedance (less than 4 ohms) when delivering maximum power.

The designer states that the most natural effect occurs when the third speaker is placed *above* the listener. This is not mandatory, however, and the speaker may be moved to a more distant, rearward location.

The control on the rear speaker lets you match the speaker to a specific room. It is a 20-ohm, 10-watt wirewound resistor with an adjustable slider. Adjust the sound level until it is not too loud to your ear.

SECTION 2

FM Stereo Reception

8

Multipath Distortion

Multipath distortion is to fm what ghosts are to tv. Each is caused by the arrival of two signals at the antenna—an original wave transmitted by the broadcast station, followed by an identical signal arriving an instant later. The second wave takes

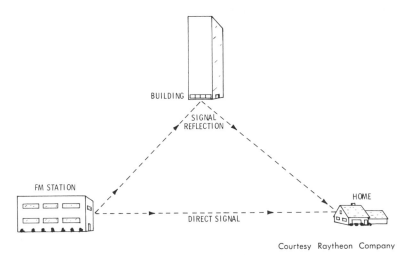

Courtesy Raytheon Company
Fig. 2-1. Multipath interference.

longer to arrive since it bounces from some surface or obstacle between the transmitter and your location (Fig. 2-1). Viewed on a TV screen, the delayed signal produces a ghost image

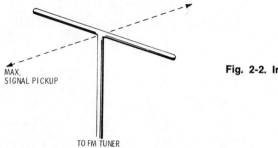

Fig. 2-2. Indoor dipole.

slightly displaced from the direct image. In fm-stereo reception, the delayed signal may create a variety of symptoms. You may hear buzzing, unusual noises, or a loss in separation during an fm-stereo broadcast.

The problem is not linked only to weak signals. Multipath distortion, in fact, is especially prevalent in larger cities where signals are strong. Tall buildings offer good reflecting surfaces for signal bounce.

The most effective cure for multipath distortion is to narrow the antenna pickup area. With sharper directional characteristics, the antenna picks up more of the direct signal and rejects multipath interference. The simplest measure occurs for an fm tuner operating on a built-in line-cord antenna. These antennas (actually a length of the ac cord which powers the turner) are notoriously unpredictable in performance. It may be possible, however, to shift the position of the cord and favor the direct signal while listening for lowest multipath distortion. If the tuner is operating on a T-shaped dipole, provided by some

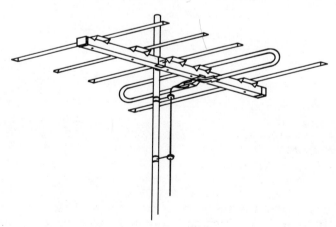

Fig. 2-3. A 5-element yagi.

manufacturers for indoor reception, attempt to orient the antenna for minimum distortion. These antennas are most sensitive when facing broadside to the arriving signal, as shown in Fig. 2-2.

When reception is on rabbit ears, attempt to angle the elements. Turning the V-shaped rods through a half circle may favor the direct signal. Also, you might change its position in the room. Listen carefully for any change in sound since very small movements of the antenna may cause a great difference.

The most elaborate method for reducing multipath reception is by erecting an outdoor fm antenna and orienting it to the

Fig. 2-4. Antenna rotator.

transmitting station. Known as a *"yagi,"* the antenna contains up to ten horizontal elements. As the number of rods increases, so does antenna sensitivity and ability to reject multipath signals (Fig. 2-3). The pitfall is that fm stations may lie at various points around the compass and no single position of a highly directional antenna will satisfy them all. This can be solved by adding a rotator (Fig. 2-4) for swinging the yagi in any desired direction. If you live in a strong signal area and the yagi antenna causes tuner overload, cure it by using the attenuator described elsewhere in this section.

9
FM From TV Antenna

There are several ways to exploit a tv antenna for fm reception. The first is a simple trick that works in some cases and costs virtually nothing. Simply connect a length of common

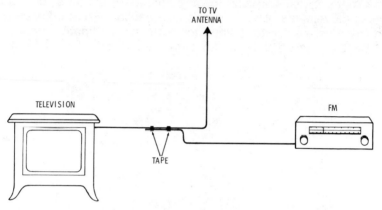

Fig. 2-5. Picking off an fm signal.

twin lead to the antenna terminals of the fm tuner. Lay the last foot or so of twin lead against the twin lead which runs from the tv antenna to the tv set. By trial and error, it's often possible to "steal" a usable fm signal from the tv lead-in. Try various points along the lead-in until you locate a position that delivers ample fm signal on stations you wish to hear. Then tape the twin leads together, as shown in Fig. 2-5. There is no direct connection; the signal transfers through electrical fields. Check to see if tv reception is disturbed by the additional twin lead, especially on channels at the higher end.

Many tv antennas cannot deliver adequate fm signals. The elements are cut to favor tv channels and there is a "hole" for fm (which lies between tv channels 6 and 7). With rising numbers of fm receivers, however, antenna manufacturers in some instances have filled the gap. If you plan to replace a tv antenna, consider a model which is designed to include the fm band, and states this coverage specifically in its literature. With a combination antenna of this type, you'll need a "splitter," like the one shown in Fig. 2-6. It divides the fm signal from the lead-in and feeds it to the fm tuner. If you use a common two-set coupler (designed for operating two television receivers from one an-

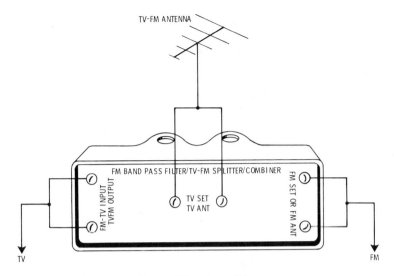

Fig. 2-6. A tv-fm splitter.

tenna), the fm tuner could suffer interference from tv channel 6, which lies close to the fm band.

Another potential fm benefit from a tv antenna is a ready-made outdoor mounting point. It's possible to fasten an fm antenna to the same mast that already supports the tv antenna. In strong signal areas, keep the fm antenna about five feet away (in the vertical direction) from the tv antenna. If you note any deterioration in tv pictures, you may have to increase spacing to 8 feet. This should eliminate interaction between the elements of both antennas.

10

Constructing an FM Dipole

If you are now receiving fm on a built-in line cord antenna, rabbit ears, or a single wire attached to the fm tuner, a folded dipole antenna may improve reception. It can be assembled for a few pennies with standard flat twin lead (300 ohms), cutters, and a soldering iron. The antenna can be used indoors or outside, and displays a figure-8, or bidirectional, pickup pattern. Arrows in the drawing (Fig. 2-7) reveal that highest sensitivity is in the two directions which lie broadside to the horizontal arms.

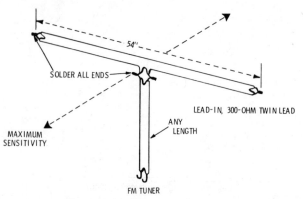

Fig. 2-7. Homemade fm dipole.

Follow the Ilustration to build the dipole, and consider these details. Start by cutting a length of twin lead 55 inches long. Strip one-half inch of plastic from the two copper wires at one end. Twist the wires together and solder them. Repeat this treatment at the opposite end of the line. This forms the top of the antenna. Fold it in half temporarily to locate the center point. With cutters, snip into *one* copper conductor at the center point. (Note that the top conductor remains unbroken.) Remove about a half-inch of plastic from the cut ends of copper at the center, and connect them to the lead-in part of the antenna, as shown. The lead-in running to the tuner may be any length.

The joint at the center of the antenna is mechanically weak, so strengthen it with a back-up plate. It must be nonmetallic. If the antenna is located indoors, ordinary cardboard, wood, or plastic will do. Tape the center of the "T" area to a back-up about two inches square. If the antenna is mounted outdoors, cover the center joint with plastic electrical tape to keep out moisture.

The antenna can provide good reception if mounted outdoors, but support all twin lead away from metal by stand-off insulators sold for tv use. Another good location for an antenna of this kind is in an attic of a wood-frame house. Recall, though, that it is bidirectional. Signals entering the ends of the antenna must be strong for satisfactory reception.

11

Better Twin Lead

Flat twin lead wire, the type for tv, is commonly used for the cable run between an outdoor fm antenna and fm tuner. It is rated at 300 ohms and it performs well in many installations at low initial cost. Common twin lead, however, is subject to aging and weather. Performance drops when it becomes wet or covered with soot, and the life span is generally considered to be from 3 to 5 years. At little extra cost, you can improve the lead-in by installing one of several superior types.

Manufacturers reduce lead-in loses by adding plastic foam between the conductors, or through forming the insulation into a hollow tube. In some cases, wires are encapsulated in a solid, low-loss material. By excluding moisture and creating a longer path for dirt to form between wires, electrical operation of a foam-filled or tubular twin lead is better than that of ordinary flat type. There is less chance for signal losses and reflections to occur along the length. Signal losses show up as increased background noises behind a station, while reflections may cause some cancellation of the stereo signal and loss of channel separation.

Even improved lines have limitations. They must be kept away from any metal by at least three inches in the run from antenna to tuner. This is done by tv stand-off insulators. More serious, in some cases, is the fact that such leads offer little immunity to electrical noise picked up along the lengths of the lines. This becomes apparent if you live near a busy highway, and ignition noises from passing cars and trucks create severe interference. In bad cases, try to relocate antenna and lead-in as far as possible from the road.

The most expensive lead-in is the shielded type. It has great resistance to dirt, moisture, and (by virtue of a metal jacket) noise pickup. Shielding also permits the line to run against metal surfaces so installation is simplified. With an electrical rating of 300 ohms, the line may be directly connected to the fm antenna and tuner. A possible disadvantage of shielded twin lead is that signal loss is somewhat greater than for other types. This shouldn't be a factor, however, unless the line is very long (say, much beyond 75 feet) and you're attempting to tune distant stations of borderline strength.

12

Coaxial Cable

Coaxial cable is the standard shielded line used in two-way radio, test equipment, and other applications. Its greatest advantage as an fm lead-in is high immunity to ignition noise and weather, and ease of installation. It may even be buried underground with no electrical effect. It is less costly than the shielded 300-ohm twin lead mentioned earlier.

Coaxial cable has two disadvantages. One is that signal loss is greater than for any of the twin-lead types already described.

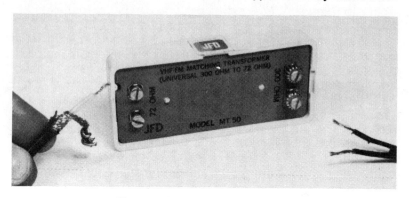

Fig. 2-8. A matching transformer.

But this should not be a major factor if the stations you receive are not steeped in background noise, and if the cable run is under about 75 feet. The other disadvantage is that coaxial cable is rated at approximately 75 ohms and therefore mis-

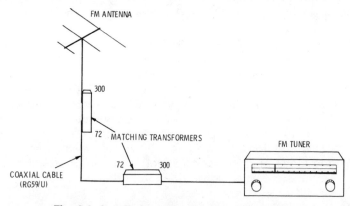

Fig. 2-9. Installation of a matching transformer.

matches the standard fm antenna and tuner input impedance of 300 ohms. To install coaxial cable, you must add special matching transformers (sometimes called "baluns") which transform the impedance, as shown in Figs. 2-8 and 2-9.

Which is the better cable, coaxial or shielded? For a new antenna installation, shielded twin lead is probably the better choice despite the somewhat higher price. It eliminates any matching problems by connecting directly to antenna and fm tuner. You should know how to handle coaxial cable, however, since it has been the standard professional lead-in for many years and is popular with many installers. You might encounter it in a cable or master antenna system, or in accessory equipment designed to connect directly to the 75-ohm coaxial line (which is typically Type RG59 U) It's simple to interconnect to this value by inserting a matching transformer wherever 300-ohm twin lead must join 75-ohm coaxial.

13

Preamplifier

A preamplifier, sometimes known as a "booster," can be added to an fm receiver to cure several problems. A booster, however, has certain limitations which should be known in advance. It cannot take a signal of poor quality, strip away the noise and deliver it static-free to the fm tuner. When a weak, noisy signal is applied to a booster, the result is a strong, but still noisy, signal. The device can only amplify what it receives. For this reason, a booster is justified mainly for the following reasons: to overcome losses in a long lead-in wire (which runs, say, over 75 feet); to increase the resistance of a lead-in to interference pickup in a noisy area; and to overcome losses caused by splitting the fm antenna signal between two or more fm receivers.

In view of the foregoing, you would not install a booster near the fm tuner to improve a noisy signal. It should mount on (or near) the fm antenna terminals. If this does not cure noise, the antenna is not delivering sufficient signal to the line, and the pickup should be improved (by a higher location, more elements, better orientation, etc.)

Do not use a cheap booster on an fm tuner of high quality. Much of the performance built into an expensive tuner goes toward achieving a high signal-to-noise ratio. Placing a poor booster in front of the tuner can degrade this quality. A good booster should also contain filter circuits which shape its response to the 88 to 108-MHz band, and thus provide protection against interference from tv channels and aircraft transmitters.

14
Front-End Overload

In most localities the listener tries to capture as much signal out of the air as possible. Yet there are instances where excessive fm signals enter the tuner and cause overload. The symptoms are distorted audio, poor stereo, and the appearance of one station at several points along the dial. These troubles may occur after you have installed a new antenna, raised your old one, or added a booster amplifier. The ability of the tuner to handle signals beyond a given threshold may have been exceeded.

A handy way to cure the problem is with an attenuator. It is a device placed in the lead-in to reduce signal strength without creating a mismatch in the line. The adjustable model is more convenient to use than the fixed type because attenuation should be applied only to the signals of stations where overloading is present. For convenience, locate the attenuator close to the fm receiver so you can make adjustments while tuning or eliminate the attenuation when it is not needed.

There are several types of attenuators available from electronic distributors. One is the continuously variable unit which turns smoothly, like a conventional volume control. It permits you to attenuate an incoming signal over a ratio of 1000 to 1. Another style is the step attenuator with a 3-position switch calibrated in dB levels. The positions may be marked "0" (no attenuation), "6" (which cuts the signal in half), and "25" which cuts the signal to 10 percent of its original value. Some experimentation reveals the best position to use.

15
FM Bandpass Filter

This is an inexpensive accessory that installs in the lead-in near the antenna terminals of the fm tuner. Consisting of coils and capacitors which tune the fm band, the fiilter can help reject signals outside the band which enter the tuner (Also see "Aircraft Interference"). The bandpass filter sharpens the front-end selectivity of the tuner and is especially helpful for rejecting "image" type interference.

It is sometimes claimed that interference from amateur and Citizens band operators is eliminated by the fm bandpass filter. This is rarely true. Transmitters in these services often cause interference by *harmonics.* These are frequency multiples which fall directly into the fm band. Filtering in these cases must be done at the transmitter, not the fm receiver. (Other cures for ham-CB interference are described elsewhere in this book.)

16

Antenna Connection

The antenna terminals on an fm tuner are generally a pair of screws which retain the lead-in wire. The wire is 300-ohm twin lead which may be connected in either direction. The lead-in is *balanced;* neither wire is considered "ground" or "hot." One advantage of this type arrangement is that it forms a low-cost cable which has some ability to reject noise without a shield. In some tuners, however, one antenna terminal is grounded inside the chassis; this can affect the antenna balance. Before permanently fastening the twin lead, tune to the weakest station you wish to hear. Touch the lead-in to the antenna terminals in either direction. Select the direction that produces better reception.

This procedure should also be tried with tuners in which neither antenna lead is connected to ground. Although there is no polarity in the system, some electrical unbalance usually occurs. This unbalance may be caused by metal surfaces near the antenna or lead-in which introduce mismatching and unbalance. Always check for the most favorable connection at the terminals.

17

Tuning Tips

It is worth the effort to develop a careful technique when tuning weak fm stations. Tuning aids (meter or magic eye) which indicate correct tuning by a *high* reading are rarely as accurate as your ear. These devices are usually inserted early in the tuner, before the detector stage, and yield incomplete information. It is in the detector, later on, that much of the signal quality is determined. A station should be centered in the detector with great accuracy by proper handling of the tuning dial.

Indicators which operate on the maximum-signal principle are correct only when the turner is in perfect alignment. This may not always be true because of poor quality control at the factory or subsequent aging of components. Further, when signals are extremely strong, meters tend to slow their response and conceal the precise tuning point. (Such indicators, however, are useful when used with an antenna rotator. As the antenna is turned, you can watch the meter and determine when the elements are pointed directly at the station.)

Fm tuners with zero-center meters—which indicate correct tuning when the needle moves to a center line—can tune most signals with great accuracy. Here, the meter is measuring zero dc voltage output of the detector, a condition that reflects the electrical balance needed to recover good audio and stereo information. During weak-signal reception, however, limited needle swing may obscure the exact reference point. Your ears should take over now as the indicator. Even on weak stations, you can often tune with great accuracy by ignoring the program material. Listen, instead, to hiss or static behind the music. As you move the dial across a station, carefully listen for least noise. (Loudest audio at this time, in fact, may be distorted.)

Some trials with this technique should prove that tuning for least noise is often better than trying for clearest audio. Further, the point where noise and static drop out is often sharp and easy to locate. If you have any difficulty hearing it, boost the treble control for a moment as you tune.

SECTION 3

Phonos and Turntables

18 Cleaning Phonograph Records

It is usually a surprise to many audiophiles that phonograph records can be immersed in water and given a bath. Not only is it an excellent technique, but the results are surprisingly good. An aging disc picks up dust, dirt, grit and fingerprint grease, and suffers increased surface noise. A record bath restores much of the original sparkle.

A few ounces of liquid detergent (like the kind used for dishes) and several quarts of water form the solution. Thoroughly wet the disc with it, then gently use a sponge, soft cloth or brush with a wiping action in the direction of the grooves. Don't press too hard. You do not want to grind any hard material into the record surface. Let the detergent solution attack and dissolve the greasy material for a few minutes. After treating both sides of the disc, flush the disc with warm, then cool tap water. Flood the water completely over the surfaces then let the discs drain. Any remaining liquid on the disc can be blotted up by paper towels or similar absorbent material.

To prevent the label from peeling away, don't return the washed record to its sleeve until it is absolutely dry. You can hasten the drying process by holding the disc over a gentle heat source; for example, about two feet above the burner of a cooking range set to low heat.

19

Speed Accuracy

The drive system in most phonographs develops extremely accurate turntable speed. This accuracy is due to the ability of the motor to lock on to the stable 60-Hz line frequency. The phono should continue to rotate at the correct speed even with 10 to 15 percent line-voltage fluctuations. Since any phono suffers friction and wear, the turntable speed should be checked occasionally.

Fig. 3-1. Checking turntable speed with a strobe disc.

Expensive phonos have a built-in strobe device to check turntable speed (rpm). Turntable speed can be checked on any machine by obtaining a strobe disc, like the one shown in Fig. 3-1, and viewing its rotation under a fluorescent light. The appropriate pattern apparently stands still if the speed is correct. If a fluorescent lamp is not available, you can obtain a neon lamp to create an identical effect (see photo). Although a conventional incandescent lamp is not supposed to create a strobe effect, somehow it does produce a faint pattern that apparently stops when the rpm is correct. The gas-type lamps, however, are much easier to view.

The strobe can help you set the speed control of a phono, if one is provided by the manufacturer. The control, sometimes found in costly equipment, allows the listener to adjust the rpm to exact musical pitch. Small speed error in nonadjustable phonos, though, shouldn't be considered serious. To meet professional standards, the number of dots on the disc passing a fixed point (your finger will do) should be about four over 10 seconds. Greater speed errors shift music pitch but it may be acceptable in normal listening.

A strobe can be used to observe what happens to the speed when you load a stack of records on the turntable, or when a dust remover is added. If they slow the turntable excessively, you might want to investigate the loss of torque in the drive system. (Some remedies are described elsewhere.)

Fig. 3-2. Remove burrs from center hole.

20
Deburr the Disc

When a record loaded on an automatic turntable fails to drop at the appointed time, it could be suffering a common affliction —a poorly formed center hole. During manufacture, the hole fails to form perfectly and it develops ragged edges. The roughness catches on the triggering mechanism in the spindle.

If there's any sign of such hang-up, the cure is simple. Use an instrument, like the pen-knife shown in Fig. 3-2, to carefully remove the burrs. Check every new disc you buy for the problem and cure it before playing.

21
Stylus Force

The pressure of the stylus on the disc should be checked regularly. Cartridge manufacturers recommend a specific force measured in grams. This value assures proper tone-arm tracking, correct compliance at the pickup cartridge, and low needle

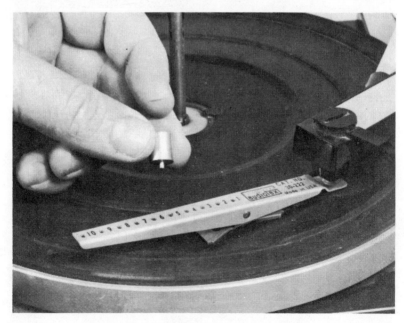

Fig. 3-3. Check for proper stylus pressure.

and record wear. Some changer mechanisms will not trip properly if the stylus force is inadequate.

Many phonos already have a built-in device for adjusting stylus force, but the measurement can easily be done with a low-cost balance scale, like the one shown in Fig. 3-3. The stylus is placed in a holder at one end of the instrument, while a movable weight is inserted into various holes until balance is achieved. The number of grams is read directly from the scale.

If your phono has a counterweight at the rear of the arm for determining stylus force, it should be checked at intervals of about six months. In systems which use a spring to determine stylus force, check at shorter intervals of about two months.

If you have a phono with a built-in stylus dial and do not have the instruction sheet, here is how the typical instrument is adjusted. First, the tone arm is balanced, the stylus dial is turned to zero, and the arm is placed in playing position near the rim of the turntable. A counterweight at the back of the arm is turned or rotated until the tone arm is perfectly balanced. It should float freely suspended in air. If there is a lock on the counterweight (a set screw or clamp), secure it in this position. Now the stylus dial can be turned to the number of grams suggested by the cartridge manufacturer.

22

Stylus Dust

In tracking miles of record grooves, the phono stylus (needle) almost certainly accumulates dust. It forms in blobs around the side of the stylus and clogs tiny, surrounding spaces. Such dirt accounts for numerous complaints of poor high-frequency response and a muffled quality to the performance. Keeping the stylus clean is simple. Do not flick dirt off with a fingertip; this could cause possible needle bending. Solvents should not be applied either.

A soft brush is the proper tool. Use it to gently loosen and remove caked and tangled particles (Fig. 3-4). For the best technique, strokes should run fore and aft (not side to side.) Camel's hair brushes are widely available for the job, with some models attaching to the phono base. Each time the arm swings toward the record, the needle passes through the brush. If this doesn't remove dust effectively, return to the old-fashiond manual method of holding the brush in your fingers.

Less dust accumulates if records are always returned to their jackets and the phono protected under a cover when not in use.

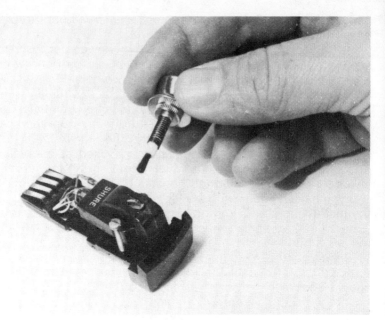

Fig. 3-4. Use a soft brush to clean the stylus.

23

Keep Phono Drive Clean

Most phonographs have simple drive systems to transfer motor torque to the turntable. Multispeed operation, (33 rpm, 45 rpm, etc.) is achieved by several diameters on the end of the motor shaft. When the operator changes turntable speed, he is actually moving a rubber wheel—the "idler"—up and down the steps of the shaft. It is like shifting gears in an automobile. The rubber idler is pressed against the inside rim of the turntable to apply driving force at the correct speed.

This system must operate without slippage. Loss of friction or tension causes two common problems—an inability to deliver correct rpm, and a stall or stoppage during a change cycle. Both difficulties can usually be cured with little effort. All it takes is removing the turntable and cleaning the affected parts.

Some turntables merely lift off after the center spindle is removed. (You may have to rotate the spindle to disengage it.) On other phonos, the turntable is held by a C-clamp. Remove the clamp with a small screwdriver twisted in the slot provided. When the turntable is accessible, turn it upside down, as shown in Fig. 3-5 and note the inner rim. It has several dark rings or markings which show the contact area of the rubber idler wheel. With a cloth dipped in alcohol, remove all traces of rubber and other foreign matter from the inner turntable rim.

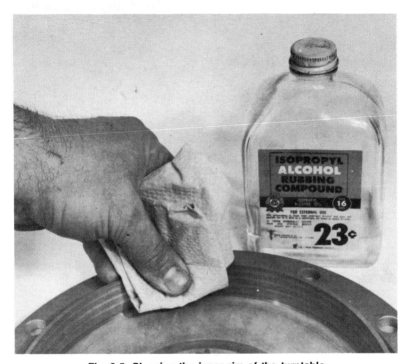

Fig. 3-5. Cleaning the inner rim of the turntable.

Next, treat the rubber idler wheel (Fig. 3-6). In some cases, merely rubbing its edge with alcohol restores the surface. Any glaze on the edge should also be removed. If the rubber shows signs of aging and loss of resiliency, replace the wheel. Suitable replacements are widely available. If a flat area is worn in the rubber rim, this is another good reason for replacement. It is caused when the operator fails to place the speed selector in a neutral position after listening to records.

Thoroughly clean the motor drive shaft. This is done by holding the rubber idler away from the shaft, and allowing the shaft to spin as you hold an alcohol-wetted clothed against it.

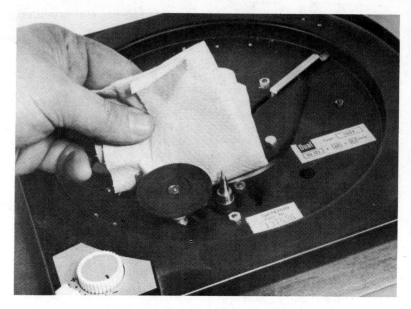

Fig. 3-6. Cleaning the idler wheel.

24
Cartridge Mount

When a cartridge is fitted to a tone arm, it must hold the stylus exactly vertical with relation to the record surface. Each wall of the groove bears one stereo channel and is canted toward the bottom of the groove at a 45-degree angle. If the pickup stylus does not point downward, making equal contact with each groove wall, stereo separation is apt to suffer.

As shown in Fig. 3-7, a simple method for checking vertical alignment is to place a thin mirror on the turntable, then placing the stylus on it. Shine plenty of light on the mirror, and align your eye to view the cartridge and its reflection. If a mounting error exists, it should immediately be apparent; any vertical line running into the reflected image will not appear straight, but bent at the mirror surface.

There is usually no provision on the cartridge for correcting such error. You can improvise an adjustment by inserting a thin washer under one cartridge-mounting screw. Install the washer needed to bring the stylus into perfect vertical alignment.

Fig. 3-7. Checking vertical alignment of stylus.

25
Turntable Lube

Slow turntable speed is sometimes traced to binding in the bearing. It can be checked by turning the speed selector to a neutral position to allow the turntable to spin freely. Rotate the turntable by hand to check for any stiffness. If there is any significant resistance, remove the turntable and clean the exposed bearing. Light machine oil (available especially for phono and recorder mechanisms) is applied to the bearing, as shown in Fig. 3-8.

On the matter of lubrication—hi-fi equipment often suffers from too much, rather than too little, oiling and greasing. The major hazard is lubricant leaking onto surfaces which must remain absolutely dry—like rubber rollers, metal drive shafts, pulleys, belts, and pressure pads. If spillage occurs, remove all traces with alcohol and a cloth. This could prevent slipping or stalling in the mechanism. If you are certain that years of operation have dried out moving parts, it is best to obtain service information for the specific phonograph (or tape recorder) and lubricate according to the manufacturer's recommendations.

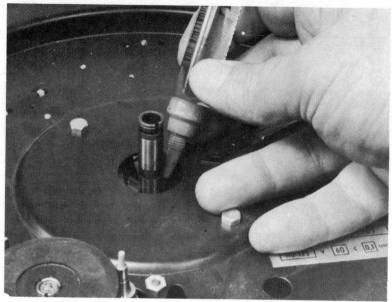

Fig. 3-8. Apply a light machine oil to the bearing.

26

Loose Belts

Signs of loose belts are often easy to detect. The phono drive slips and fails to maintain proper speed. In some instances, speed may be correct until the turntable is loaded with records or dragged by a dust brush mounted on the tone arm. In any case, belts may be expected to stretch and relax after several years of service. Close inspection of the mechanism while turning belt-driven parts by hand usually exposes the problem.

Replacing the belts is the best treatment. For this reason, it is a good idea to order a spare set or two at the time of the original purchase. These items are sometimes difficult to secure a few years after the introduction of a specific model. (Some manufacturers of excellent products have gone out of business, which can permanently disable your equipment.) If belt slippage is still minor, or renewal impossible, try a "nonslip" compound. As shown in Fig. 3-9, some of this fluid dabbed on a belt may improve and extend its gripping power.

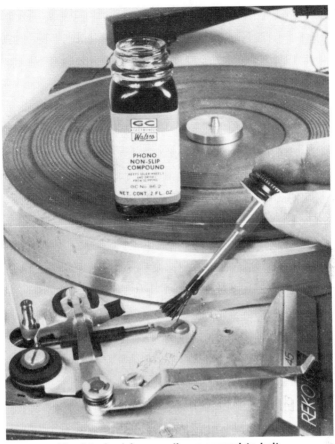

Fig. 3-9. Applying nonslip compound to belt.

27

Master Shut-Off

A feature provided in some automatic phonos is a master shut-off switch. After the last record ends, the tone arm returns to the rest position and switches off line voltage to the complete system. This is useful for late-night operation (especially when the listener falls asleep and leaves equipment on), or when an amplifier is located away from the listening area. The automatic switch may be available as an option from the original phono manufacturer, or as an accessory from others. Check to see if it will work with your specific model.

Fig. 3-10. Semiautomatic timer.

Another automatic shut-off is a simple, spring-wound timer, like the device shown in Fig. 3-10. It is not completely automatic; you preset its knob to switch off power any time up to four hours. This is useful for fm listening only. If a timer controls a tape recorder or phonograph, it could switch off ac power in the middle of a selection. This would leave pressure rollers engaged and possibly cause a flat spot to develop in the rubber.

28
Leveling the Turntable Base

Use a small level to check a turntable base. Round bubble levels are especially made for the job, but you can also do it with the level shown in Fig. 3-11. If you use the straight type, as shown, be sure to check for level in both directions. Unless the phono manufacturer provides some method for leveling the turntable, fasten thin wood shims under the base, as needed.

Fig. 3-11. Check to see if turntable is level.

A level condition aids the stylus in contacting walls of the record grooves with equal force. This is important for good stereo. Another factor affected by turntable leveling is the ability of a stylus to properly track. With very light pickups and loud audio, it's easy for a stylus to leave the groove and skid across an out-of-level disc.

SECTION 4

Amplifiers

29
Quiet Location

Turn the input selector of your amplifier to magnetic phono and listen, with no program playing. Since this position operates at high gain, it is particularly susceptible to stray electrical fields that generate hum. If you can hear hum with your ear positioned less than about two feet from the speaker, the preamplifier stage may be immersed in a powerful electrical field. This can be demonstrated by moving the amplifier from its location and listening for changes in hum level.

Transformers, motors, and other electrical equipment might be the source. Although an amplifier is shielded against hum, the circuit might not have sufficient resistance against concentrated fields.

One pitfall in a separate component system occurs when a preamplifier is located next to the power transformer of a main amplifier. The electromagnetic radiation from the transformer may not be completely excluded by the preamp cabinet. Rearranging or relocating equipment might solve the problem. In some cases, small shifts in position can significantly reduce hum levels.

30
VU Meter

A meter connected directly to an amplifier output can help balance stereo channels, or give a continuous reading of relative output power. The meter is an ac voltmeter calibrated in "VU" and connected to the speaker terminals. Two such meters are shown in the amplifier of Fig. 4-1, each monitoring one stereo channel. (The amplifier is a home-made circuit, which explains the missing labels and dial markings.) The meters were added after the amplifier was completed, and the same can be done to commercially wired units.

Fig. 4-1. Amplifier with two VU meters.

What does a VU meter read? The letters mean "Volume Units" and they refer to an exact electrical standard that enables broadcast stations to interconnect with the telephone company or feed a known signal to a line. The reading of "0" on a VU meter refers to an audio signal of 1 milliwatt into a standard 600-ohm line. Since a home hi-fi system has no 600-ohm line, VU becomes a simple, relative reading of amplifier output.

Do not attempt to connect the VU meter across some point in the circuit other than the speaker terminals. The meter movement may load the grid of a tube, for example, and short-circuit the signal. Some VU meters have an internal resistance of about 8000 ohms which could degrade a circuit with an impedance of more than about 1000 ohms. Connecting the meter to the low impedance of speaker terminals should produce no ill effects.

There is another factor to consider. Since the VU meter is meant for a 600-ohm line, you'll have to calibrate the meter for your amplifier. It can be done by wiring a potentiometer in series with the meter, playing the amplifier at a high listening level, then adjusting the potentiometer for a "0" indication. Use a carbon potentiometer of about 100,000 ohms or more, and the highest impedance output on the amplifier (usually 16 ohms). This should be done with a steady tone feeding the amplifier, or the VU needle will bounce and be difficult to adjust. Note that the "0" position is slightly more than half-way up the meter face and allows higher audio peaks to register.

If other instructions are provided with a VU meter, follow the manufacturer's recommendation. Some dual meter units are available with built-in calibrators if you do not want to assemble your own.

VU meters also have a 0-100–percent scale. This is ignored in the home since it's a carry-over from broadcast station practice. When a signal of "0" is sent over the line from a studio console, it modulates the radio transmitter at 100 percent, the maximum value allowed.

31

Dummy Load

The hobbyist who tinkers with audio amplifiers should find this item invaluable. To check the output or frequency response of an amplifier, for example, you place a meter across the speaker terminals, introduce a test signal, and run the volume wide open.

You would be deafened by the noise, and maybe burn out the speaker voice coils. Also, speakers are rated at only an approximate number of ohms (it varies with frequency) and your measurements would prove inaccurate anyway.

It is vastly better to disconnect loudspeakers and substitute a dummy load, like the one shown in Fig. 4-2. It absorbs audio power in utter silence and converts the energy to heat. It is an accurate load, too, since it is a pure resistance and it presents the same number of ohms to the amplifier regardless of frequency. This is important if you wish to measure actual power output of an amplifier, as detailed in the next item. Another advantage is cost.

The load is made from nichrome wire sold in hardware stores to replace the burned-out element in an electric toaster or

Fig. 4-2. Dummy speaker load.

broiler. The one shown in Fig. 4-2 is a typical replacement with a power rating of 660 watts (far greater than the power of most audio amplifiers).

To use nichrome wire as a load, stretch out its coils slightly and connect one end to the probe of an ohmmeter. The meter should be on a low resistance scale (usually R × 1.) Move the other probe along the turns of nichrome until you read the number of ohms you want for the load, e.g., 4, 8 or 16 (the whole length measures about 20 ohms.) Add a half-inch for connecting to the speaker terminals, then snip off the remaining wire.

Two precautions: If you run tests with an amplifier putting out high power, the nichrome wire becomes hot, so keep it from touching anything flammable. Also, the wire should not be permitted to contact other metal and become short-circuited; it changes the load. (There's little voltage hazard, however, since the audio voltage is low, despite high power.)

When a resistance wire, like nichrome, heats up, its resistance increases. In tests using this dummy load, however, the resistance change was not sufficient to affect measurements.

32
Voltmeter Checks Watts

It is a distinct advantage to read the actual number of watts generated at the speaker terminals of an amplifier. A meter reveals losses in power that may go entirely unnoticed by the ear. Thus, an occasional power check may warn of impending trouble. In stereo systems, you can quickly determine volume or balance settings that produce equal power in both amplifiers. You can observe the dynamic relationship between stereo channels; vary the audio output of one channel while reading steady power of the other. Any interaction may indicate the inability of the power supply to completely energize both channels at the same time.

An audio wattmeter measures the output of an amplifier, but the instrument is expensive and mainly used by service technicians. You can take the same readings at virtually no cost with the common voltmeter popular among experimenters. It may be a vom, multimeter, vtvm or solid-state voltmeter. All it needs is a scale which measures *ac* volts from 0 to about 50.

Two other items are required: an accurate dummy load and a source of steady tone. The load is constructed by using a length of nichrome wire, as described elsewhere in this section. You might, for example, cut it to exactly 8 ohms and connect it across the speaker terminals of the amplifier. Steady tone is secured from an audio oscillator or a test record. All that's needed is several seconds of a 400-Hz (or similar) tone to make the meter indicate. The tone should be a sine wave, since other waveforms might make the ac meter readings incorrect.

You can measure maximum output of the amplifier by placing the meter probes across the dummy load (Fig. 4-3) while tone is playing and the volume control turned fully on. (Tone controls should be at the zero, or flat, position.) Watch for the

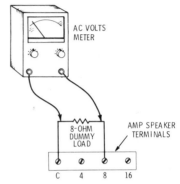

Fig. 4-3. Measuring amplifier watts with a voltmeter.

reading on the ac meter. In this example, 20 volts means the amplifier is putting out 50 watts. Finding the value is done by converting volts to watts with simple arithmetic. Since the load resistance is known (8 ohms, in the example) it's easy to apply Ohm's law to convert volts (E) into watts (P).

$$P = \frac{E^2}{R}$$

$$P = \frac{400}{8}$$

$$P = 50 \text{ watts}$$

Rather than work out this formula every time, it is easier to make up a chart that converts ac voltage to watts at about a half dozen power levels; say for 1, 2, 4, 8, etc. watts.

This test, incidentally, assumes that your ac voltmeter is indicating an *rms* (root mean square) value. This is true in most instruments, but check the meter face or instruction manual to verify it. Some meters read the ac signal in *peak* volts (but may include an rms scale, as well.) If you can measure only peak ac readings, convert them to rms by multiplying by .707. For example: a 14-volt peak indication is equivalent to 9.9 volts rms. Thus 9.9 would be the figure to insert in the formula given above.

33

Frequency Response

Manufacturers publish frequency response curves which detail the performance of hi-fi amplifiers. These specifications are highly controversial. Even the experts do not often agree on the conditions under which measurements should be taken. Some authorities declare that frequency response is meaningless when taken at low power levels, while others maintain that wide frequency response is not as important as the degree of distortion generated by the amplifier. The frequency response argument will not be pursued here—with one exception. If you can record frequency response curves when the amplifier is operating perfectly, you will have an excellent reference for

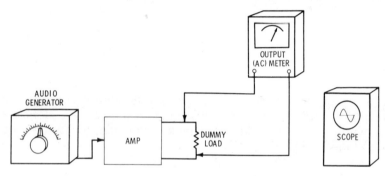

(A) Bench setup.

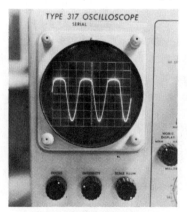

(B) Oscilloscope display with flattened peaks.

(C) Display with perfect sine wave.

Fig. 4-4. Setup and equipment for

detecting future trouble. Curves give a sweeping view of how the amplifier performs under dynamic conditions. If you suspect that something is amiss you can recheck your data to determine whether the amplifier or some other component is causing trouble.

Some basic equipment is needed to perform the frequency response test: a dummy load connected across the speaker terminals (covered earlier in this section); a meter which can read ac volts, with a decibel calibration (commonly found on multimeters and vtvms); and an audio generator (Fig. 4-4E) to introduce test tones into an auxiliary or tuner jack. An oscilloscope is invaluable to visually monitor the output signal across the dummy load. (Speakers are not used since their volume would prove unbearable.) The bench setup is shown in Fig. 4-4A.

Turn everything on and adjust tone controls to their flat position. With a 1000-Hz tone applied to a high-level input (e.g. auxiliary), turn up the amplifier volume control while observing

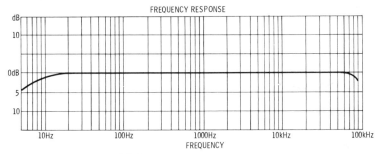

(D) Frequency response curve.

(E) Audio generator.

(F) Turn bass and treble fully on.

checking frequency response.

the signal on the oscilloscope. Increase the signal until you just see signs of flattening in the wave on the screen, as shown in Fig. 4-4B. (This should be at approximately the rated power output of the amplifier.) Back off on the volume until you see a perfect sine wave on the screen (Fig. 4-4C) (you can compare the shape with the input signal from the generator).The reading on the output meter is now considered "0" dB and serves as the reference.

A frequency response chart of a commercial circuit is reproduced in Fig. 4-4D to serve as a guide. Feed a succession of tones into the amplifier, starting at about 20 Hz and continuing up until output drops off sharply (above 20 kHz). Using our graph in Fig. 4-4D as a guide, note the dB reading for two dozen or more frequencies throughout the audio spectrum. When the points are joined by a line, you'll have a frequency response curve like the one pictured, with the number of dB "down" or "up" (cut or boost) from the reference line at 0. Note that a 3-dB drop cuts power in half, while a 3-dB boost is equal to doubling in power. But these large changes in amplifier wattage will be barely heard because of the special (logarithmic) response of the human ear.

Once you have plotted the curve, repeat it with bass and treble controls in full boost position (Fig. 4-4F), then, plot a third curve with tone controls fully off. The resulting traces will reveal tone-control action over a complete range. It takes a few hours to develop all this data, but you'll have an important profile on how your amplifier functions.

One precaution while taking the output (dB) readings—the input signal from the audio generator must be constant for all frequencies. If your generator has no built-in meter to monitor level, verify it with the ac output meter whenever you change frequencies. The input tones must be identical in strength throughout the complete test range or the curves will be incorrect.

34

Electrolytics

Electrolytic capacitors, usually encased in metal cans (Fig. 4-5), are prone to more problems than many other amplifier components. They serve as filters in the power supply and are a frequent cause of hum. Electrolytics lose their effectiveness in time through increasing electrical leakage or an open-circuit condition. The underside of the capacitor may be seen to ooze

Fig. 4-5. Electrolytic capacitors.

a powdery chemical, but this is not a positive sign of trouble.

A quick test that gives an approximate idea of the condition of the electrolytic is to turn the amplifier volume control down to zero. With tone controls in their normal position, you should hear little or no hum from a loudspeaker, with your ear positioned about three feet away. Electrolytic cans should not be uncomfortably warm to the touch, either.

A trick by technicians to check an electrolytic is to bridge across a suspicious unit with a good one. Observe correct polarity, and note that not all such capacitors have their cases, or negative side, connected to a chassis ground. If hum level disappears during this test, it is advisable to replace the capacitor.

The rectifier tube or silicon diodes in the power supply should also be checked at this time. In some cases, a shorted power diode applies ac to an electrolytic capacitor and causes damage.

SECTION 5

Tape Equipment

35
Less Print-Through

Print-through is a tape problem that's heard as a ghostly echo during playback. The cause is easy to trace. The magnetic field on one layer of tape passes to an adjacent layer and transfers (or "prints through") part of the program. There are several cures, and the ones to use depend on how valuable you consider a particular recording.

The first step is not only the easiest, but the most important for minimizing print-through. Do not record at excessively high levels. In turning the recorder gain control up into distortion, or maximum, region, the chances for print-through are multiplied. If your tape equipment is of high quality, and it produces little noise at low recording levels, then reduced recording gain can also help the problem.

Another solution is to use special, low print-through tape. It might be worth the extra cost if you intend to store important tapes for long periods.

It is well to remember that even regular tapes are less susceptible to print-through if they are occasionally played. Playing tends to break up layers that remain stored in direct contact. A final tip is used by some broadcast technicians. They never rewind a tape immediately after playing. Fast motion of the transport tends to pack tape layers tightly together and storing tape in this condition favors print-through. It is prevented by rewinding the reel immediately before the next playback.

36

Nice Splice

Some handy recordists can take ordinary scissors and splicing tape and neatly join two strips of magnetic tape. But for most of us, some special technique and mechanical assistance make the job easier. They eliminate the two pitfalls of poor splice—an annoying "plop," and a snag as the spliced tape rides over the recorder heads.

Never use ordinary cellophane tape. It will run and ooze between tape layers and stick them together. Avoid magnetized

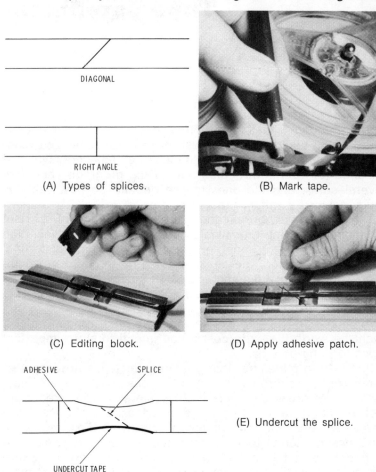

Fig. 5-1. Splicing recording tape.

tools in cutting the tape. A razor or scissors could record a pop on the tape as you cut. If you can pick up a pin with your cutting tool, demagnetize the tool (using a tape head demagnetizer).

Tape can be joined with two types of cuts (Fig. 5-1A) so you'll have to choose between them.

Diagonal—This is the usual splice, achieved by cutting the tape ends at a 45-degree angle. The advantage of this cut occurs when different tapes (meaning those with different background noise levels) are spliced. The cut slides gradually over the playback head and you will hear a blend between noise levels. There is no abrupt transistion.

Right Angle—In this splice, tape ends are cut squarely. The benefit here is that you lose the least amount of program material. This is important when editing is close—when attempting to edit out a single click, syllable, or other brief interval. When using the right-angle cut, be sure to leave no space between the tape ends. It causes a click on playback.

To find the exact point of a splice on the uncut tape, audition the tape, then stop the unwanted bit just before it rides across the playback head. It may require a little practice, but it is not difficult. (Try to remove the head cover if you can.) Use a grease pencil ("China Marker" sold in stationery stores) to make a dot where you will cut (Fig. 5-1B).

Fig. 5-2. "Gibson girl" semiautomatic splicer.

Once the undesired tape is excised, use an editing block (Fig. 5-1C) to make the splice. The block acts as a jig and aligns the tapes perfectly for making the cut and pressing on the adhesive patch. Do not overlap the tape ends. They must butt together with no intervening space. The adhesive tape (Fig. 5-1D) is applied only to the shiny side of the tape, then pressed into a firm bond by rubbing with a fingernail.

Final step is to remove the sticky edges of the adhesive tape to prevent snagging in the machine. Use scissors to undercut the tape (Fig. 5-1E). Or, if you want to do the job semiautomatically, use a "Gibson girl" cutter, like the one shown in Fig. 5-2. This device is more convenient than the simpler editing block and delivers just the right amount of indentation. (It's called "Gibson girl" after tight-waisted beauties of the 1920s.) The spliced tape is also tight-waisted at the undercut.

37

Broken Tape

For tapes that have accidentally broken, use a butt splice to disturb the recording as little as possible. The ragged ends (Fig. 5-3) are directly butted together, using the editing block shown earlier. There should be absolutely no space or overlap at the joined ends. Undercut the splice as already described.

The major problem with broken tape is the stretch often suffered just before the break occurs. It creates a "wow" in the program that cannot be eliminated by the butt splice. If wow is too objectionable, you may have to remove a stretched section of tape and sacrifice a bit of program material.

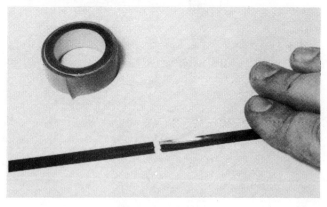

Fig. 5-3. Broken tape.

38

Clean Heads

Even tiny deposits on tape heads produce vast differences in recorder performance. Tape sheds oxide particles as it passes through heads and the accumulation prevents the tape from making intimate contact. It causes loss of high frequencies, muffled or squealing sounds, or changes in playing speed. Since head cleaning is so important, virtually all equipment manufacturers allow quick access to heads, usually through a removable cover. Check the instruction manual for details on your model.

Apply the recommended cleaner (which often may be ordinary rubbing alcohol) on a cotton swab or "Q-tip." Carbon tetrachloride was once popular for cleaning, but avoid using it. The fumes can prove dangerous and the liquid mars some finishes. As shown in Fig. 5-4, the cleaner is applied directly to the shiny finish of the head. While you're at it, remove any oxide or dirt from any item along the route of the tape (like metal posts or guides).

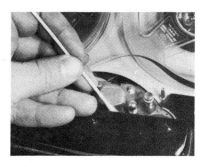

Fig. 5-4. Cleaning the heads.

How often do you clean heads? About every four weeks should be adequate. It is an easy job that has a significant effect on sound. Clean the heads more often if you are a heavy tape listener.

39

Degaussing

If a tape recorder is operated in perfect electrical balance, its metal parts would never become magnetized. Various signals —record, playback, and bias—alternate between plus and minus, which produces a *net* magnetizing force of zero. But

Fig. 5-5. Degaussing a tape transport.

recorders, like most man-made devices, are far from perfect and unsymmetrical electrical signals tend to magnetize parts in the transport mechanism. The result is an increase in noise and hiss on both playback and record. Fortunately, with a degauseer it is easy to rid a transport of residual magnetism in moments.

The job should be done about once a month. If you're planning to record an important program, or play back a valuable tape, first degauss the machine. Otherwise, you might cause tape deterioration in both cases. The principle of the degausser shown in action in Fig. 5-5, is this: A thick coil of wire draws heavy current from the ac line and converts it into powerful magnetic fields. Since the fields alternate 60 times per second, metal molecules on the transport are jumbled out of their magnetized state.

But the degausser can also inflict harm if not used correctly. If you turn it off while holding it close to the recorder, you could stop the action at a critical instant and worsen the condition. The proper technique is remarkably easy: Start by holding the degausser at least three feet from the transport and plug it in. *Slowly* walk toward the machine and hold the degausser tip very close to the various heads (don't scratch them). After several seconds, move over other metal parts, like guides and posts. After you've demagnetized them, *slowly* retreat. Wait until you're at least three feet away from the deck before unplugging the degausser.

40

Mechanical Noises

When a tape transport makes mechanical sounds that originate inside the cabinet, a frequent cause is the ventilating fan.

As shown in Fig. 5-6, the fan is mounted on the shaft of the drive motor and spinning blades may strike nearby cables. It not only produces noise, but a possible slow-down in tape speed.

Fig. 5-6. Noise could be caused by ventilating fan.

The problem may occur because of poor lead dress and anchoring, and loose or bent fan blades. Signal cables with excess slack may be at fault, but more likely it is the ac connections to the motor. These are two black wires seen in the photo running to the motor at the lower right. Use plastic electrical tape to hold these wires in a safe position. While you are checking, be certain no cables have shifted position and are resting against the hot glass of a vacuum tube.

41

Pressure Pads

Wear and aging in the felt pads which press tape against recorder heads take their toll. The tape may audibly squeal as pads lose resiliency and fail to dampen vibration. Another symptom is uneven head wear. You may see this by closely examining the head under a bright light; the wear areas are not outlined by sharp vertical lines. Defective pads also cause poor tape-to-head contact and loss of audio quality. Speed variations, including wow and flutter, are sometimes traced to pressure pads.

The first step in treating pressure pads is to clean them regularly. All traces of oxide or other material is removed with a "Q-tip" moistened with water. If the pads are hardened and smooth, you might get additional life out of them by scraping away the surface glaze.

The pads, however, ultimately wear out and require renewal. There are two approaches. The easiest method is to obtain exact replacement pads from an electronic distributor or the recorder manufacturer. The pads are often supplied mounted on arms and replacement is a simple matter. If pads alone are provided, remove the old ones with a razor blade and scrape away all traces of cement. Glue the new pads in place with a small amount of cement and avoid any spillage into the deck.

If you cannot locate exact replacement pads, fashion your own from felt material (Fig. 5-7) sold for the purpose at electronic dealers. These are supplied in two thicknesses—1/8-in. and 3/16-in.—to cover most models. Choose the one that most closely matches your machine and cut the pads to size. They have a self-adhesive backing so no cement is required for mounting on the arms.

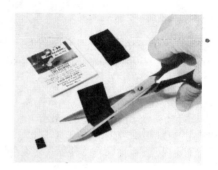

Fig. 5-7. Pressure pads.

42
Replacing Heads

The first sign of worn tape heads is usually a loss in high tones. Wear is caused mostly through abrasive action of tape moving over the heads. Audio losses occur as the heads lose metal and tiny head gaps become enlarged. (High frequency response is directly related to the width of the head gap.) Wear tends to be uneven, which also prevents an intimate contact between the tape coating and head.

Be cautious before replacing recorder heads. There should be no question about their worn-out condition. You can avoid unnecessary trouble and expense if the actual fault is dirt, misalignment, incorrectly adjusted recording bias, or some other defect. Replacing recorder heads is not as simple as changing a phonograph stylus and the job shouldn't be attempted unless you have specific service literature provided by the manufacturer. Small installation errors can lead to serious losses in performance. Certain items of test equipment might be required (like an alignment tape or voltmeter.) Despite all these precautions, a hobbyist with some ability in electronic troubleshooting should be able to do the job.

If new heads are not available from the original manufacturer, you can obtain them for almost any machine through electronic dealers. (One producer specializes in supplying replacements for about 2000 different models.) It is important to get heads as close to the original electrical and mechanical specifications as possible (Fig. 5-8). If you use replacement types, chances are that small modifications might be required, like washers, shims,

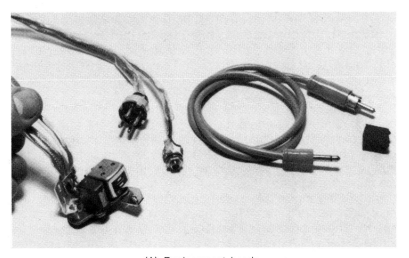

(A) Replacement heads.

(B) Removing old heads.   (C) Replacements installed.

Fig. 5-8. Changing heads.

or a new connector or cable. With care, the differences shouldn't prove difficult to handle. After the installation is done, align the new heads as described elsewhere in this section.

43
Tape Speed

A simple cardboard strobe disc placed on a turntable measures the speed of a phonograph. To determine the speed of tape, however, measurement is done with a more elaborate instrument, but one that also operates on the strobe principle. When the markings on a rotating wheel (Fig. 5-9) appear to be stationary, tape speed is correct. The one shown here measures two popular speeds 3¾ in./s and 7½ in./s. It is viewed most clearly under a fluorescent light.

Absolute speed accuracy is not critical in much home recording. Since the same error affects both playback and record, there may be no audible difference. Long-term changes in recorder speed, however, can make early recordings made on the machine sound too slow on playback. If tapes from another machine are played, the speed error may also introduce annoying differences. Most disturbing of all is the attempt to play commercially recorded tapes on a machine which is off speed. These problems should be visible on the tape strobe since it derives its reference from the very accurate line frequency (60 Hz) supplied to the home by the power company.

Speed problems—usually slowing of the transport with age—generally point to aging in the drive system. Rubber rollers, clutches, belts, and other components which mechanically operate the recorder should be checked for wear and replaced.

The tape strobe is intended to check only open-reel decks since cartridges, cassettes, and similarly sealed packages may render the tape inaccessible to the strobe wheel. A section of tape must be available to place against the wheel for motive

Fig. 5-9. Checking tape speed.

power. The strobe should not induce excessive drag and slow the transport.

On some recorders, usually of very low-cost manufacture, tape speed varies with the amount of tape on the reel. The strobe cannot operate on this type. Most home recorders, however, deliver constant tape speed through a rubber roller which presses the tape against a turning metal shaft.

Head Alignment

It is impossible to achieve good high frequency response without careful head alignment to insure that the head gaps lie at right angles to the direction of tape movement. Called "azimuth" adjustment, it's described in the service literature for the machine. If loss of highs is severe and you are sure the transport is free of foreign matter, realignment might be appropriate.

One precaution though, you might deteriorate the quality of tapes recorded earlier on the machine. Yet, if you are convinced that high-frequency loss is excessive, or that you want the recorder to operate with maximum quality on prerecorded tapes, then head realignment is desirable.

You should have the instructions provided by the manufacturer. They show where the adjustments are for vertical height and azimuth alignment. Head height should be checked to assure that each track is fully covered by the heads. Azimuth is adjusted by using a specially prerecorded tape. An output or ac voltmeter is used to monitor the effect of head position as you do the job. You will be adjusting the head for maximum re-

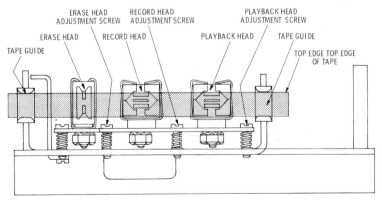

Courtesy KLH

Fig. 5-10. Typical head adjustments.

sponse to the high-frequency test signal recorded on the tape (Fig. 5-10).

45

Excessive Hum

When hum during playback and record is objectionable, a common cause is a strong magnetic field induced in the heads. As described elsewhere for magnetic phono cartridges, tape heads contain a fine coil of wire that is susceptible to electromagnetic radiation.

Two sources are the power transformer in a nearby amplifier, and the radiation of a fluorescent lamp. Try moving the tape transport away from these sources to minimize any pickup. Often, a few feet produces satisfactory improvement.

46

Wow and Flutter

These speed irregularities are often caused by problems in the drive system. First, check the rubber pressure roller. It is usually accessible after removing the head cover (Fig. 5-11).

Fig. 5-11. Pressure roller accessible with cover removed.

Any sign of oil on the rubber must be removed with alcohol. If the rubber has become hardened, replace the roller. The wheel is easy to take out by removing a C-clip or screw. When the surface of old rubber has become glazed, it is sometimes possible to remove the shine with fine sandpaper. This restores "tooth" to the rubber and improves its grip on the tape. The rubber, however, should be replaced if it has lost resiliency.

Another source of slippage and speed error is loss of pressure between the rubber roller and capstan. This is often corrected by replacing the spring which controls the roller. Cases of flutter are sometimes traced to a defective flywheel bearing. When wow and flutter tend to occur during the second half of the tape reel, there is probably too much hold-back tension acting on the feed reel. This may be a problem of lubrication, but consult the manufacturer's service literature before attempting the cure.

Rubber Belts

One of the most annoying problems that can befall a tape recorder is a broken belt. Although cost is low, it can disable a machine for weeks, or perhaps permanently. This is often true if the machine is an import that is several years old. Because of model changes and the difficulty of contacting the manufacturer, a replacement belt may not be available. Some electronic distributors carry replacements, but too often they fail to list hundreds of specific models. But don't despair. Although a belt is almost certain to break or stretch in time, you can try "replacing the replacement."

Bring the broken belt to an electronics dealer. In consulting his catalog he may discover that no replacement is available for your model. The catalog, however, often states major dimensions (circumference and inside diameter), as well as a drawing of the belt. Choose the one that most closely matches yours. Note that belts are made in round and square (cross-section) styles.

If the old belt is not available, you can still attempt to figure out the replacement. Run a string around the two pulleys in the recorder which hold the belt. This provides a circumference dimension. The groove on the pulley can be measured as a guide to belt thickness.

SECTION 6

Loudspeakers

48
Mounting Board

Fastening a loudspeaker to a board can cause distortion if not done correctly. The speaker has a delicate mechanical suspension that is held within a few thousandths of an inch tolerance by the manufacturer. A warped board or overtightened mounting nuts can upset the tolerances.

Check a board with a straightedge before mounting the speaker. It should show no curve. Next, place the speaker over the screws on the board with extreme care. This operation has ruined many an expensive speaker—the installer fails to match the holes in the speaker frame with the mounting screws, pushes forward and punctures the paper cone. Hold your fingers temporarily *between* the speaker frame and mounting board and don't push until you are absolutely certain the frame holes are aligned with the screws.

Tightening the speaker to the board is another critical step. It is possible to tighten the nuts unevenly and exert tremendous leverage that warps the speaker frame. Try it this way: After the speaker is slid onto its mounting screws, press the frame firmly against the board to take up any slack. Hold it there. Tighten the nuts by finger pressure only. The final tightening is done by using a suitable tool to rotate the nuts an equal number of additional turns. Do not turn the nuts until they seize in position. They should be firm, not locked.

49

Speaker Wires

Ordinary lamp cord, the kind used for appliances, is popular cable for connecting loudspeakers. It is No. 18 stranded wire. Where a speaker line is to be concealed under a carpet, flat tv twin lead can usually serve. Twin lead conductors are often No. 20 stranded copper. For typical wiring runs in the home, say under 50 feet, these cables are sufficiently large in diameter to prevent undue loss of amplifier power.

If you want to run extension speakers, however, the matter of wire size could become important. Some wires sold for speaker use are simply too thin to deliver full power over great distances. You can easily determine the smallest wire size for a speaker line by using Table 6-1. It is based on a permissible power loss of 15 percent. Even the most sensitive ears will not detect this loss and the figure is considered acceptable for professional public-address systems.

Table 6-1. Wire Size and Speaker Impedance

Wire Size	Speaker Impedance (Ohms)			
	2	4	8	16
No. 10	150'	300'	600'	1200'
No. 12	95'	190'	380'	760'
No. 14	60'	120'	240'	475'
No. 16	38'	75'	150'	300'
No. 18	23'	47'	95'	190'
No. 20	15'	30'	60'	118'
No. 22	9'	18'	37'	75'

Note that wire size depends on another factor besides distance. The impedance of the line also affects power. (Impedance refers to the speaker and amplifier output values—typically 4, 8, or 16 ohms.) As impedance drops, so does the permissible length of speaker line. For example, if you use regular lamp cord (No. 18) you may run an 8-ohm speaker up to 95 feet away from the amplifier, without any serious loss. But use a 4-ohm speaker and the maximum distance for No. 18 wire drops to 47 feet. The table should enable you to come up with any possible run in the average home.

When wiring several speakers on the same line, be aware that the total impedance becomes that of the speaker *combina-*

tion. For example, two 8-ohm speakers wired in parallel makes the line 4 ohms. Wired in series, these speakers make the line 16 ohms. The total figures are used when consulting the table. (Further details on speaker impedances in various combinations are discussed in another section.)

50
Extension Speakers (Parallel)

The simplest method for adding an extension speaker is the parallel hookup, as shown in Fig. 6-1. If both main and remote speakers are 8 ohms, the parallel circuit reduces the total to 4 ohms. You connect the line to the 4-ohm output on the amplifier. (The diagram is for mono; use two remote speakers for stereo and the identical hookup.)

The disadvantage of a parallel arrangement is that it drops the line impedance. As shown earlier in the speaker-line chart, this limits the maximum line length. It could be a problem, say, when attempting to run a speaker to a patio 50 or more feet away from the amplifier.

In any parallel arrangement, if both speakers have the same impedance, the amplifier simultaneously delivers the same amount of power to each one. If this causes excessive volume at the main speaker use one of the level control methods described later.

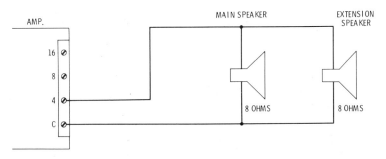

Fig. 6-1. Hookup of speakers in parallel.

51
Extension Speakers (Series)

An advantage of the series hookup between main and remote speakers, shown in Fig. 6-2, is that the total line impedance is raised. Since the impedance of series speakers adds, two 4-ohm units total 8 ohms. The higher impedance permits longer wiring runs, as shown in the speaker wiring chart given earlier.

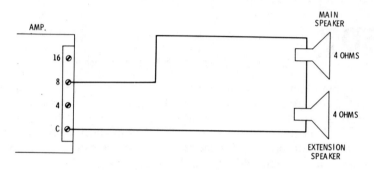

Fig. 6-2. Hookup of speakers in series.

The disadvantage of a series arrangement is that some interaction can occur between main and remote speakers. This may be unacceptable if the main speaker is a very costly instrument and the distant speaker a small, inexpensive unit. You can avoid the problem by using a changeover switch (described elsewhere) to choose either main or remote operation.

52

Long-Distance Speaker

There are certain situations that call for an extremely long run of wire for an extension speaker—from a summer home out to the beach, or between two points hundreds of feet apart. Normal wiring techniques fail because the amplifier power is lost in the resistance of the line. As our speaker-wire table (given earlier) reveals, an 8-ohm speaker running 600 feet from the amplifier requires a No. 10 wire size. This creates the ridiculous combination of huge conductors (like those which bring electrical power into the home) and very high cost. It would take a couple of men to lift the wire anyway.

The simple, inexpensive solution is using two audio line transformers to raise the impedance of the line to 500 ohms. This vastly reduces the wire requirement; you can now run tiny No. 22 wire more than 700 feet with negligible loss. The hookup, shown in Fig. 6-3, uses a pair of "line-to-voice coil" transformers, commonly available in radio parts catalogs. The Stancor

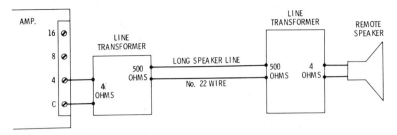

Fig. 6-3. Using line transformers to reduce line loss.

A-7947, for example, is rated at 500 ohms on one side, and 6 to 8 or 3.2 ohms on the voice coil side.

In using these line transformers, do not expect true hi-fi performance. It is assumed they will operate with extension speakers where high quality is less important. Also, be sure to remain within the wattage rating of the transformer, which normally is a few watts, but this should be enough to drive a small remote speaker. A line transformer of wide frequency and high wattage would be prohibitively expensive in this application.

53 Main-Remote Selector

Some audiophiles prefer to keep their expensive speaker systems as pure as possible. They want no interaction from remote or extension speakers tied to the same line. Yet, they'd like to enjoy the low-cost method of hearing stereo in other rooms through extension speakers. This is easily solved by a selector switch installed at the amplifier. It permits the operator to send audio to either the main or remote speakers (Fig. 6-4).

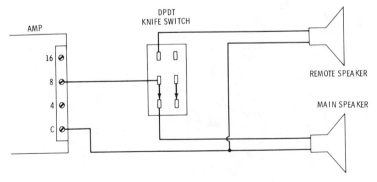

Fig. 6-4. Wiring the main-remote switch (one channel only shown).

Unlike earlier hookups shown, there is no simultaneous sound in main and remote areas; it is either one or the other.

A good choice for the selector switch is an ordinary bakelite knife switch commonly sold for antenna changeover. Almost any dpdt (double-pole double-throw) switch, however, will work. If the system is mono, wire to only one side of the switch, as shown in Fig. 6-4. For stereo use both sides and wire them the same way.

It is assumed that both main and remote speakers will have the same impedance rating. Thus, there'll be no mismatch when the selector is turned to either system. One important precaution—never operate the selector while the amplifier is playing. The momentary loss of the speaker load could cause damage to some solid-state amplifiers. Do any switching while the amplifier is turned off.

54

Remote Volume Control

In operating a remote speaker, you may encounter a problem in volume control, especially if speakers are located in several areas. Different speaker sizes, efficiencies or room acoustics could introduce annoying differences in level. One way to solve the problem is installing an individual speaker control.

A good device for the job is the L-pad (Fig. 6-5). It is a double-element, wirewound control which changes speaker volume without causing interaction with other speakers sharing the same line or amplifier. You can even turn off a remote speaker with an L-pad, but first consider some precautions before using this type of control.

Fig. 6-5. L-pad volume control.

The L-pad can only *reduce* power. It is passive and cannot amplify. For this reason, it shouldn't be used to correct a condition where a speaker near an amplifier is too loud, while distant speakers play too softly. This is caused by either a serious mismatch in the line or cables that are too thin to carry the power. It should be corrected by consulting the speaker wire chart or one of the speaker hookups shown elsewhere.

Since the L-pad operates by converting unwanted audio power into heat, the control continues to absorb wattage from the amplifier regardless of volume heard in the speaker. Even when the controlled speaker is turned off, the pad draws the same power. This might be a consideration if amplifier power is on the low side (say, 10 watts) and your main speakers are low-efficiency types. The pads and remote speakers could draw enough continuous power to be a problem. One answer is to add a main-remote switch, shown in an earlier item, to keep all remote operations off the line while listening to main speakers.

Wire an L-pad according to the diagram in Fig. 6-6 if the manufacturer supplies no instructions. Note that two center terminals of the unit are jumpered together with a short wire. The pad itself can be conveniently mounted on the remote speaker cabinet. If the cabinet is mounted above convenient reach, insert the pad anywhere along the line.

The impedance (or ohms) of a pad should be the same as that of the speaker; that is, an 8-ohm speaker takes an 8-ohm pad. It is possible, however, to use one pad for two speakers, if both speakers are in the same room. (We don't mean two speakers for left and right stereo channels. These must be handled by individual controls to preserve stereo separation.) Two speakers can be controlled by one pad if you follow the series and parallel hookups already shown. For example, one 8-ohm pad can simultaneously control two 4-ohm speakers connected in series. One 4-ohm pad can control two 8-ohm speakers in parallel.

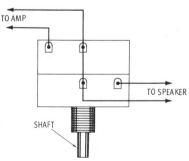

Fig. 6-6. Wiring on L-pad volume control.

55
Remote Stereo Balance Control

Here is a dilemma you might encounter in extending stereo to an additional room in the home. While listening in the main area, you are able to adjust audio balance in the main speakers by a control on the stereo amplifier. But due to differences in acoustics or other factors, the same setting doesn't produce good stereo at the remote speakers. One technique is to install a 100-ohm potentiometer at the remote speaker that sounds louder when you play a mono program through the system. The control permits you to reduce its sound to bring about a balance between the two speakers.

The control must be a wirewound type. It assures sufficient wattage capacity to handle the audio power. Do not use a simple control of this type to control main or high-power speakers. It is intended for small inexpensive units. Fig. 6-7 shows how the control is wired.

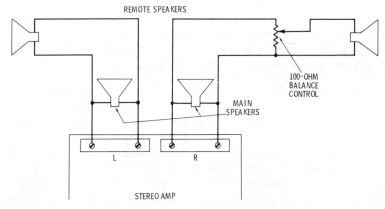

Fig. 6-7. Remote stereo balance control.

56
Speaker Switching

This is another approach to controlling sound in a speaker. Instead of a gradual change in level, the switch turns the speaker completely on or off. This is usually the approach in commercial switching units sold for hi-fi speaker control. The idea is that a switch disconnects the speaker from the amplifier line and substitutes a fixed resistor in its place. The resistor, actually a dummy load, converts the power into heat. If you do not wish to purchase a commercially made unit, you can easily assemble one to meet your requirements.

As shown in Fig. 6-8, the switch is a single-pole, double-throw (spdt) type. It can be a slide switch so long as it can handle about five watts of power. In this arrangement, a mono speaker is shown being controlled, but it is easily converted to stereo

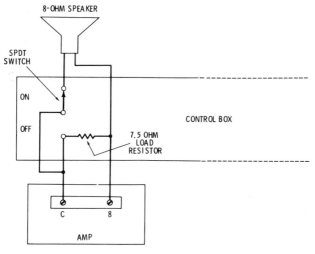

Fig. 6-8. Speaker switching hookup.

by using a dpdt switch. If you use the switch with a solid-state amplifier, obtain a nonshorting type to prevent possible damage to the transistors. As an extra precaution you can turn down the amplifier volume just before switching a speaker in or out.

The load resistor is selected for a resistance that is closest to the impedance of the speaker. If the speaker to be switched is in a low-power system, then a carbon resistor of 2-watt rating should serve. You can get these resistors in 7.5, 8.2 ohms and other values. If your amplifier is in the high-power category, it is safer to use wirewound resistors of 10-watt rating. They are available in 4-, 7.5- and 15-ohm values, close enough to match any amplifier output.

57

Impedance Selector

A difficulty in wiring extension speakers is maintaining proper impedance matching. In systems described earlier in this secton, it was assumed that speaker combinations added up to the right number of ohms for connecting to the amplifier. This is not always possible to achieve and it causes power loss and unbalance. You can, however, install a switch at the amplifier output that allows rapid choice of any impedance (Fig. 6-9). For example, if the main speaker is rated at 8 ohms, the switch is turned to that position while listening in the main area. When

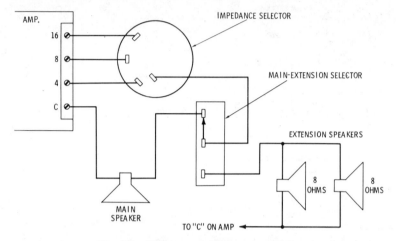

Fig. 6-9. Adding an impedance selector.

listening on extension speakers, the selector is turned to another impedance if necessary. In our diagram, there are two 8-ohm extension speakers wired in parallel for a total of 4 ohms. The impedance selector is instantly turned to that position. Other combinations are also possible and you avoid rewiring the amplifier terminals each time the speaker system is altered.

Note that a main-extension selector is also used. This is similar to the main-remote switch described earlier and is a spdt switch for choosing either system. If the system is stereo, then use a dpdt switch and wire both halves of the switch in the same manner.

The impedance selector in the diagram is shown wired for mono. The switch is a 1-pole 3-position rotary type, nonshorting. If you want to perform the same function for stereo, use a 2-pole 3-position switch. A Mallory Type 3223J should meet the requirements.

58

Torn Paper Cone

Rips in a paper cone can occur because of an accident during speaker installation. Another possibility is a child inserting a sharp object through the speaker grille. The symptom of a torn speaker is fuzzy sound; the edges of tear rub together. Distorted audio may not occur on all frequencies; but it may be triggered when the program material contains certain frequencies. In most instances, the repair is easily done if the right materials are available.

Do not use a cement that shrinks. It could cause wrinkling in the speaker cone. A special compound for the job, usually known as "service cement" is widely available from electronics distributors. Another choice is rubber cement sold in local stores which carry stationery supplies. If you have any doubt about the cement, tear a piece of writing paper for several inches, then join the ragged edges with cement. After it is dry, check for any wrinkling or distortion in the paper.

Fig. 6-10. Cementing a tear in a speaker cone.

In most instances, joining the torn edges of the speaker is enough to stop any rubbing action. A small puncture (made by a pencil, for example) should be sealed by forming a bridge of cement over it. Only when an opening is large, should a patch be applied. Make the patch out of paper and attach it according to the cement-manufacturer's instructions for using his product. If the cone is shredded beyond repair, check with the speaker manufacturer. In many instances, he can recone the speaker.

59
Gritty Voice Coil

This is a case of preventive medicine since a gritty voice coil is much easier to avoid than to repair. The symptom is similar to that of a torn speaker—fuzzy or distorted sound when music plays certain tonal combinations. Close inspection of the speaker, though, reveals no breaks in the paper. The trouble is revealed by gently pressing on the paper cone, near the center (as shown in Fig. 6-11). As your finger moves you should feel no rubbing sensation, just a slight tension as you push. In some cases, a scratchy vibration is picked up by the paper cone and mechanically amplified as you push the cone in and out. There are two culprits in the case of the rubbing voice coil.

One is physical distortion introduced when the speaker was originally mounted on its board, as covered in an earlier item. The other is foreign material lodged in the narrow region between the voice coil and an internal magnetic pole piece. The sealed cabinet of many speakers prevents entry of grit, but where a raw speaker is in a reflex-type cabinet or in a wall, it may enjoy less protection, especially in handling. The author

Fig. 6-11. Checking for a rubbing voice coil.

once had an expensive coaxial speaker playing in a home workshop. The back of its cabinet was left off for several days and speaker performance commenced to deteriorate. When the speaker was sent back to the factory, a technician discovered that the space between voice coil and magnet was glutted with steel particles. Somehow, an electric drill used near the speaker dispersed metal particles which ultimately found their way into the speaker. The powerful speaker magnet held them tightly in place. So keep raw speakers covered. The voice coil repair can be made only at the factory.

60
Correcting a Cheap Enclosure

The reputable loudspeaker manufacturer takes great pains in the design of his speaker enclosure. Dimensions, materials, and construction techniques are carefully selected. Yet, some excellent speakers are sometimes mounted in enclosures which ruin their performance. It may happen when a carpenter assembles the speaker cabinet as part of a console that contains the complete hi-fi system. He may not be aware of the factors that affect speaker performance. Sometimes a good speaker is mounted in a simple box for reasons of economy. Whatever the reason, consider some enclosure basics that contribute to good speaker performance. You might be able to improve the sound of a low-cost cabinet with one or more of them.

The enclosure must not vibrate. If it does, stiffen each wall with a long wood brace screwed to the inside surface. The strips may be ¾-inch wood stock. Also, install strips where cabinet walls join, using both glue and nails (or screws) to keep everything rigid. The walls of any speaker cabinet should be at least ¾″ thick. For holding on the back cover, there should be a wood screw at least every 4 inches, and one in each corner. If you hear buzzes, rattles, or other vibrations that are not in the music, chances are the cabinet construction needs more attention.

The inside of an enclosure requires acoustic damping material to deaden resonances. On each inside surface, glue plenty of fiberglass batting (1-inch thick) or other material sold for the purpose. If the enclosure is very small, insert enough fiberglass to fill the whole interior.

If the grille cloth in front of the speaker is not made especially for hi-fi remove it. Soft cloth can absorb high frequencies. Suitable grille material of wide mesh is available.

These steps will produce no dramatic improvement if the cabinet is extremely small. Simple hi-fi enclosures (a sealed box) need about five cubic feet to obtain good sound from a 12-inch speaker. A minimum size, for example would be 2 ft. wide × 2½ ft. high × 1 ft. deep.

SECTION 7

Headphones and Mikes

61
Choosing Headphones

Listening through headphones can multiply the delights of a stereo system at little extra investment. You can enjoy music when others are sleeping, run the volume at any level and play any kind of music. The sound quality of headphones does not equal that of a full-size speaker system, but the effect is impressive and enjoyable. Room acoustics cannot interfere with the performance as with speakers, and stereo separation through phones is exceptionally good. Before buying a set of headphones consider several major points.

Poorly designed phones quickly become uncomfortable to wear. It is not only weight that counts, but how the pressure is applied. Ear pads must distribute weight evenly and the phones should have an easily adjustable head band. Almost any headphone is comfortable to wear for a few minutes but it is the longer period that is important.

Phones must form a good seal with the ears (but without creating painful pressure). Unless phones are closely coupled, low tones are lost from the music. A good seal, though, may cause an unpleasant heating around the ears, so check this point too. A popular material to prevent build-up of pressure and heat is the foam cushion.

In judging earphone performance, expect good sound, but not room-shaking bass. When overdriven by low notes, earphones produce distortion. At the upper end of the audio spectrum, listen for distinct high tones without a "peaky" or sharp-edged quality that can soon grow tiresome. Listen for smooth response over the audible range and use your subjective judgment for determining which phones sound best.

Most headphones operate on the "dynamic" principle (similar to the electromagnetic coil and permanent magnet in a loudspeaker). More costly instruments are "electrostatics," which require additional components for operation. Because of the complexity of electrostatics, they should be expected to deliver very smooth frequency response over a wide audio range.

62

Add a Headphone Jack

Many stereo amplifiers are fitted with a jack to accommodate low-impedance headphones from about 4 to 16 ohms. This feature can be added with the simple accessory shown in Fig. 7-1. It requires three basic parts—a jack, a selector switch, and a pair of fixed carbon resistors. These components may be installed on the existing amplifier if the hobbyist wishes to drill holes. An alternative is to use a small metal box for housing the parts.

The device operates in two ways: you may hear the speakers in usual fashion, or switch the selector to silence the speakers and divert audio to the headphones. The 3-circuit jack is a

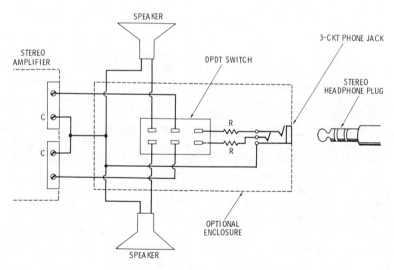

Fig. 7-1. Adding a headphone jack.

standard type which accepts the phone plug provided with most stereo headphones.

Consider some details on parts and construction. The dpdt (double-pole double-throw) switch is the loudspeaker-headphone selector. It may be a toggle or slide type, so long as the contacts can carry about 2 amperes or more (voltage rating is not important).

The two components marked "R" in the circuit are fixed carbon resistors of half-watt power rating. The resistance of these units will depend on your particular amplifier. A typical value is 100 ohms. This should introduce enough resistance to prevent the phones from receiving excessive power and blasting in your ear. If, after the circuit, is completed, you find that earphone level is still too high, then add an additional 100 ohms in series with each resistor. The ideal resistance allows the volume control on the amplifier to remain at approximately the same position—normal listening level—regardless of whether you are on phones or speaker. Headphone level and stereo balance to suit individual taste are still determined by the amplifier controls.

The dotted line shown around the circuit indicates the optional enclosure for the various components. The circuit wires may be No. 18 between the amplifier and box, and No. 18 or 20 inside it. The phone plug at the far right does not have to be purchased since it's usually supplied with the typical stereo headphone.

Home-Made Stereo Headphones

Cost shouldn't stand in the way of hours of pleasurable, exciting listening on stereo headphones. A couple of items from a local electronics store and a pair of low-cost miniature speakers can be assembled into excellent phones (Fig. 7-2). Once you have located the materials, it takes only a few hours to build them at a fraction of the price of commercial equivalents.

Most important are those tiny speakers. They are widely available as replacements for transistor portables, and should have a diameter of about 2 inches. The ones shown in the illustrations are 2¼ inches in diameter and 8-ohm impedance. You don't need those exact values but they agree with the instructions and dimensions given here. If your units are slightly different, it shouldn't be difficult to modify construction accordingly.

The perforated board on which the speakers are mounted is

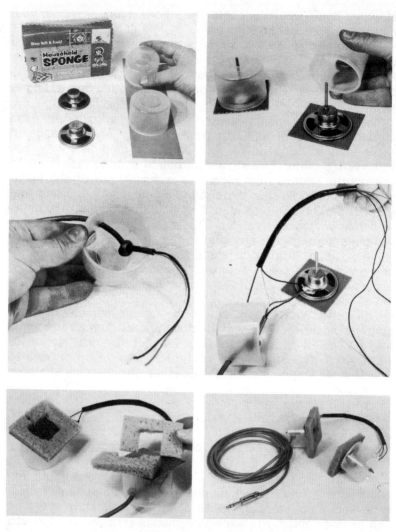

Fig. 7-2. Constructing a stereo headset.

the phenolic type sold for experimenter circuits. Any sheet plastic can be used if small holes are drilled to allow sound to pass. Carefully apply a thin ribbon of glue to the raised cardboard edge of the speaker and press it to the perforated boards.

The cups which enclose the back of the speakers are plastic caps taken from the top of aerosol spray cans. They are 2½ inches in diameter, but other plastic cups of that approximate size should serve as well. The cup is held to the perforated board with a 6-32 × 1¼-in. long machine screw. Note that the head of the screw is attached to the back of the speaker. If the metal will take solder, then solder the screwhead to the speaker frame; otherwise, fasten it with epoxy glue.

The cable is a 3-wire type (No. 22 stranded conductors) inserted into the cup, as shown. Note that a knot is tied in the cable inside the cup for a strain relief. Your wire colors may not be the same as those shown in the diagrams (Fig. 7-3), so follow the same general circuit. Final connections are determined later.

The ear pads are simply a household sponge (the soft type) cut with a scissors into half-inch thicknesses and glued to the perforated board. They provide an air seal that is essential for good bass reproduction.

The headband is a piece of semiflexible No. 18 copper wire, but you may substitute similar material (like a coat hanger). Do not use soft copper sold for bus bar. The wire should be hard-drawn for good springiness. One end of the wire is pushed into the plastic cup, then given a right-angle bend with pliers to retain it. By pushing the wire slowly through the plastic, you can get a friction fit that enables you to adjust headband size by pushing the wire in and out of the hole. The length of the headband wire from cup to cup is determined by putting the assembly on your head and checking for a loose fit. (The wire in the original model measured 14½ inches.) Final adjustment is done later.

Use two stranded wires to carry the leads over to the second speaker. They are inserted through a piece of rubber or plastic tubing slid over the headband wire for wearing comfort. Note that the wires enter each cup through a small notch cut into the plastic. Allow some slack wire to permit easy disassembly at a future time.

The cable which runs from the cups to the phone jack is about 10 feet long. This permits you to sit well away from your stereo amplifier, if necessary. The plug at the cable end is a standard 3-circuit type that matches the phone jack on most stereo equipment. (If your amplifier has no such provision, it can be added, as described in another section.)

After the project is wired and ear pads glued on, put on the headphones and adjust the headband by bending. You want a good ear seal and comfortable fit. If the band is too large, carefully push one of the copper wires further into the cup. Trim the headband wire if it is too long.

Finally, listen to program material whose left and right channels are definitely known (this where a stereo test record is

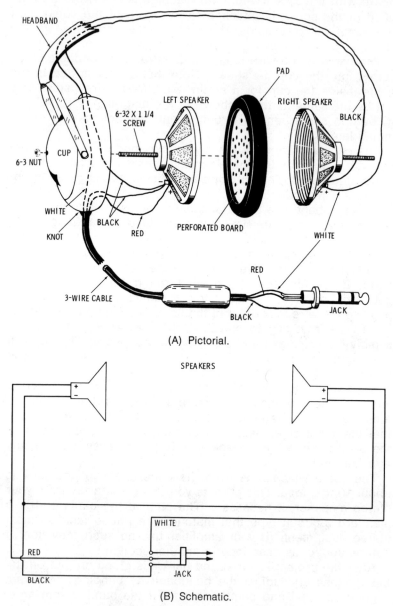

(A) Pictorial.

(B) Schematic.

Fig. 7-3. Wiring the stereo headset.

handy). If channels are reversed in the phones, unscrew the phone jack and reverse the "hot" (red and white) leads. The black, or common, lead remains the same. Another improvement might occur if the phones are phased. If there is no phasing switch on your amplifier, open one cup and experimentally reverse the leads to one miniature speaker. This is done while listening to material rich in low tones. Swapping the leads to one speaker may improve bass response in some cases.

64
Long Mike Cables

Many tape recorders are supplied with high-impedance microphones. The impedance is determined by the nature of the element, usually crystal or ceramic. A restriction on such mikes is cable length. A high-impedance line is very susceptible to hum pickup and loss of high frequencies, when extended more than about 10 to 20 feet. The distance is ample for many home recording situations, but falls short when working with large groups or halls.

The technique used by professionals is to operate the line at a low impedance. Lines rated this way may run several hundred feet, with great resistance to hum and treble losses. The amateur can do the same thing. But first, consider the two alternatives. One is to purchase a pair of microphone transformers. These convert, or transform, impedance from low to high, or vice versa. (They work in either direction.) One transformer is placed at the microphone end, and one at the recorder. The line between now operates at a low impedance.

This approach, though, may be false economy. The cost of two mike transformers may exceed that of a new mike that is not only of a higher quality, but capable of either high- or low-impedance operation. It has a selector switch or an internal reconnection for choosing impedance. This eliminates one mike transformer since the microphone end of the line needs no impedance step down. One transformer is located at the tape recorder input. So, compare the cost of a transformer against that of a mike.

The second consideration is that the low-impedance microphone most likely is a dynamic type of superior quality. High-impedance mikes furnished with many recorders are best reserved for speech and other narrow-range sounds. The dynamic mike should do a better job of recording instrumental music or choral singing when recorded over long cables.

Check your recorder literature for the microphone impedance ratings. High impedance is generally considered to run from about 10,000 to 50,000 ohms. Low impedance is 50 to 600 ohms. Fig. 7-4 shows how a dynamic mike might run through a long cable to the high impedance input on a recorder.

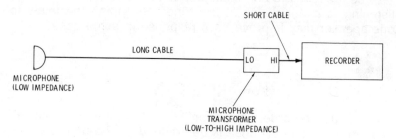

Fig. 7-4. Running long mike cables.

65

Choosing a Mike Pattern

Three basic patterns are offered by microphone manufacturers, and you might wish to pick one to improve your recordings. As shown in Fig. 7-5, they are omnidirectional, bidirectional, and unidirectional. Consider their relative merits.

The omnidirectional mike (also called nondirectional) is equally sensitive to sound arriving from all directions. You can group several people or instruments around it or place it in the center of a table for a conference-type pickup. The disadvantage of the omnidirectional mike is that it cannot reject unwanted sound.

The bidirectional mike has a figure-8 pickup and finds greatest application in broadcast stations. Since it favors only two directions, it is usually used for the face-to-face interview. But its special pickup pattern also makes the bidirectional mike least common of all.

Most popular is the unidirectional mike, also called the cardioid (because of a heart-shaped pattern). The relatively dead region opposite the sensitive side gives the recordist a chance to manipulate the pattern to reject unwanted sound. For example, he can pick up an orchestra on one side, while reducing distracting audience noises on the other side. If the performance is also going out over a public-address system, the dead side of the mike reduces the tendency of feedback. Though successful mike placement takes some experience, the cardioid pattern is the most versatile of the three types.

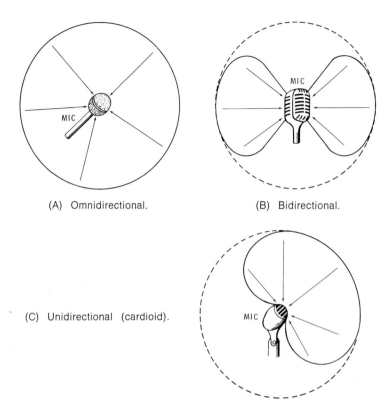

(A) Omnidirectional.

(B) Bidirectional.

(C) Unidirectional (cardioid).

Fig. 7-5. Microphone patterns.

66

Extra Mike

The principle of a dynamic microphone is that of a moving coil within the poles of a permanent magnet. As the coil moves in the magnetic field, electrical currents are induced. This is similar to the operating principle of a loudspeaker. For this reason, it is possible to reverse the operation of a loudspeaker and have it serve as an extra or spare microphone at very little cost. Instead of listening to the speaker, you talk into it. (The technique is used in intercoms.)

The parts required, as shown in Fig. 7-6, include a small speaker and an audio transformer. A handy speaker size is the 2-inch type widely available as a replacement for a transistor portable radio. Its impedance is not important. The transformer

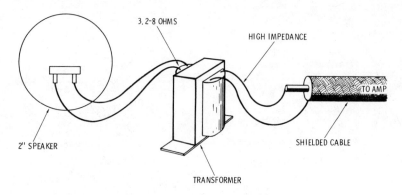

Fig. 7-6. An extra "microphone."

is an inexpensive miniature audio type. The rating is not critical, but the low-impedance side should be approximately the impedance of the speaker. The other winding of the transformer should have an impedance of several thousand ohms or more. A small audio-output transformer from an old radio or tv should also work.

Mount the "mike" inside a small enclosure, with perforations to let the sound in. (If the mike is used without a case, its gain and frequency response suffer.) A small metal "minibox" will offer some protection against hum pickup. Choose one large enough to contain the speaker and transformer. The high-impedance side of the transformer connects to a shielded cable which runs to the amplifier. (Ground the shield to the metal case.) Fit the amplifier end of the mike cable with a suitable plug for mating with a mike or low-level input. Like the dynamic mike which it simulates, this mike must feed a preamplifier stage to achieve sufficient gain.

67

Mike Noises

You have probably seen them on tv newscasts—a cover which fits over a microphone for outdoor audio pickups. If you intend to try outdoor recording, consider one of these windscreens. They reduce the rumbling sound produced by wind passing over the sensitive grille of a mike. On a windy day, such noise can obliterate the sounds you wish to record.

The wind screen, made of a foamy material, may be designed by the manufacturer for your mike and offered as an optional accessory. Since the material is flexible, another model may easily stretch to fit on your mike. These screens, incidentally, can also reduce a tendency toward "popping" in some mikes. These are annoying explosive noises produced when the speaker utters certain sounds (notably the letter "p").

SECTION 8

Techniques

68
Free White Noise

One engineering textbook defines noise as a random signal which interferes with a desired signal. There are about a dozen sources, from cosmic rays arriving from distant galaxies to electronic agitation inside a circuit. A special quality of noise is that it exists on no single frequency, but is a collection of many audio or radio waves.

Engineers painstakingly try to reduce noise, but with one notable exception. Noise is a broad-spectrum signal which makes it an excellent tool for testing. You can have all you want—free of charge—for valuable tests in the home. You find it merely by adjusting an fm tuner between stations. (If the tuner has an interstation noise suppressor, turn it off.)

That familiar sound of frying eggs and escaping steam is *white* noise. It consists of audio of virtually every tone (much as white light is a combination of all colors). Although noise through an fm tuner may sound predominantly high pitched or hissy, it contains low and midrange frequencies. What's it good for?

By tuning white noise on a stereo system you can make quick comparisons on balance and quality between the two channels. Adjust various controls—tone, volume, balance, etc.—until white noise heard in each speaker is as closely matched as possible. (Be sure the system is adjusted for monophonic operation at this time.) If you run out of adjustment possibilities, and there is still a significant difference in sound, try a change in speaker locations by experimentation. Once you have matched the channels as closely as you can, mark control positions with an identifying dot. The marks let you to return to the balanced condition at a future time.

69
Reversing AC Plugs

In many installations, simply reversing the ac plug in the wall outlet reduces hum level heard in the loudspeakers. Although the line doesn't have steady polarity—it reverses 60 times per second—one side is directly connected to the electrical ground of the building (usually water pipes or a copper rod driven into the ground). This provides a favorable path for certain circulating hum currents in equipment chassis.

There are usually several ac plugs in a typical hi-fi system—amplifier, tuner, phono, or tape deck. Since they all may affect overall hum level, each must be experimentally turned in its socket. Repeat the process several times to determine the combination that produces least hum. The job should be done while the amplifier is turned to a magnetic phono input (no program playing) with volume and bass turned to higher than normal listening level.

70
Cable Coding

If you own a component-type system, you probably have a tangled nest of wires behind the amplifier cabinet—several shielded cables, speaker wires, ac line cords, ground leads, and the like. When the amplifier or some other item is removed for servicing, it may take a considerable amount of time to locate all wires and return them to their proper connections. It is much easier to code each cable. The hookup goes faster and error is completely avoided.

Almost any kind of adhesive tape you can write on (masking, fabric, frosty cellulose) will do. Cut off a 2-inch strip, fold it in half and sandwich the cable you wish to mark between the halves (Fig. 8-1). These "flags" are located several inches from the end of the cable. Write a number on the cable flag, and the identical number on a piece of tape pressed near the matching terminals. For example; code a speaker wire and its amplifier connection with "1," a phono cable and its input socket with "2," etc. If you want to be sure that individual wires of a pair

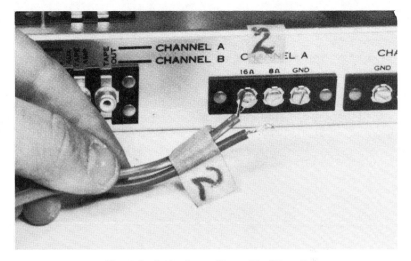

Fig. 8-1. Code the cables with "flags."

are reconnected in the original hookup, split the pair for several inches and flag each wire. This would be helpful where you determined correct speaker phasing and wanted to preserve the exact connections.

71
Assembling Phono Plugs

The standard connection between hi-fi components is through a shielded cable which terminates in phono plugs. The cables are available preassembled with molded plugs and ready for use. Since cables often develop troublesome connections at or near the plug, you should know how to make the repair. The technique of attaching a new phono plug on a cable is also used to make your own cable assemblies in any quantity or special length.

The dimensions given in the accompanying illustration (Fig. 8-2) are approximately correct, but check them against your plugs since small variations might occur. Start by stripping away an inch of outer jacket from the end of the phone cable (Step 1). Use a sharp knife or razor to slit the jacket, but don't penetrate far enough to nick the shield wires underneath. Now unwrap the shield (Step 2) and twist its wires together. If you do not have a phono cable with a spirally wrapped shield (the easiest kind to handle), use a pointed instrument to unbraid the

95

fine shield wires. Once the shield is away from the cable and twisted, apply an iron to the end and flow on a bit of solder to tin the shield.

Carefully remove the insulation from the inner wire (Step 3) leaving ¼-in., as shown. Tin the end of the inner wire with the least amount of heat and solder you can apply. Insert the inner lead into the phono plug. You should be able to push the lead until it appears at the plug tip. Push the cable into the plug until the inner insulator butts against the eyelet inside.

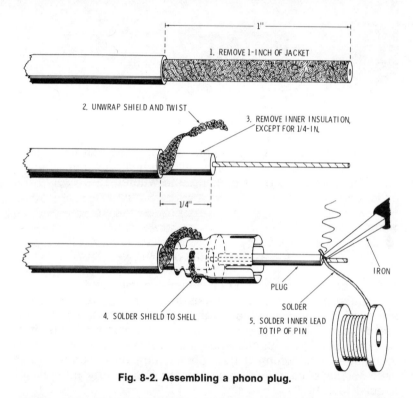

Fig. 8-2. Assembling a phono plug.

Wrap the shield around the outside of plug and solder it directly to the shell (Step 4). Be sure to touch an iron to the shell, but remove heat as soon as solder flows. Allow the plug to cool before continuing. Next, touch the iron to the plug tip *and* the inner lead protruding from the tip. Let them heat for a moment, then touch solder to the opening in the tip and allow solder to be drawn into the hole. Do not use more heat than is needed to melt the solder, then remove the iron quickly. You do not want the inner insulation to melt and cause internal short circuits. Trim away any excess shield and the extra bit of wire protruding from the plug tip.

72

Trade-In

It has been said that the quickest way to improve a hi-fi system is to trade it in for a new one! There is a demand for old, but serviceable, equipment from people who are seeking a bargain, those who wish a second set for the home, and people looking for an inexpensive system for a vacation home. A classified ad in the local paper or a trip to an audio dealer may bring results.

The price of used equipment is subject to tremendous variation. Certain name-brands command high prices regardless of age, while others are nearly worthless soon after the initial purchase. Fast-changing fashions in hi-fi may cause mahogany speaker enclosures to zoom one year, then give way to teak wood the next. A dealer might be overstocked in the item you want to sell—or he may be looking for the very unit you're offering.

Some shops or mail-order dealers will display your equipment on consignment and extract a 20 or 25 percent commission on the sale. You set the price, but some professional advice is helpful to avoid excesses in either direction.

What can you expect to realize from used equipment? If your instrument is reasonably popular, and not affected by some caprice of the marketplace, there is a simple system for determining what a dealer might offer you. Find the age of your instrument in the first column, then multiply its original list price with the factor shown in the second column:

Age (years)		Multiplier
1		.36
2		.28
3	(example)	.23
4		.18
5		.15

In the example, a three-year old fm tuner which carried a list price of $95 might bring a trade-in of $21.85. It is the policy of some dealers to apply this sum only against the purchase of new equipment. And take a tip from the used car merchants—shine the equipment until it gleams.

73

Give It More Air

Tubes, transistors and most other electronic components last longer with good ventilation. Consider several pitfalls in circuit heating, and what you can do about them.

First, allow the circuit to breathe. It may not be apparent to the eye, but most home electronic equipment cools by thermal convection. Heat developed in the chassis causes air to rise and escape upward. Cool air, meanwhile, is pulled in at the bottom to flow over the parts. Anything which interferes with the cycle raises temperature and shortens component life. So be certain that no openings in equipment cabinets are restricted or blocked. Allow at least an inch clearance from a wall for air circulation. This applies to solid-state rigs as well as tube types. Transistors, with no glowing filaments, generate far less heat than do tubes, but semiconductors are responsive to temperature. As heat increases, the transistor conducts extra current, which turns into heat, more current, and so on. It may destroy the transistor through "thermal runaway."

If hi-fi components are mounted in a custom-made cabinet, be sure provision is made for convection cooling. The cabinet-maker may produce a work of art, but one that is hostile to equipment life. One way to assure a path for convection currents is to locate holes or screened openings in the bottom of the cabinet. The back of the cabinet should have plenty of openings, as well. This technique has been used successfully in millions of tv receivers; cool air is pulled in through the bottom, while heated air is launched out the rear.

Use good judgment when stacking equipment. Placing a preamplifier atop a power amplifier may be a neat, space-saving arrangement, but it may also bathe the preamp in damaging heat. And, do not place anything electronic atop a heating unit, like a radiator.

Forced-air ventilation is one answer to cramped, overheated hi-fi. (Vacuum tubes have been known to last twice as long when operated in a stream of fan-driven air.) You can rig a fan of your own by consulting the electronic catalogs for a unit made for the purpose. Its special quality is low-noise operation and least disturbance to the musical performance.

74

Switching Surges

Some radio-broadcast stations never turn off certain pieces of electronic equipment. The reason is that a change in operating temperature varies the tuning accuracy of a critical instrument like a frequency monitor. A subsidiary benefit in continuous operation is that components, especially vacuum tubes, are less prone to failure. It would seem that continuously energized equipment would burn out sooner, but there is another factor at work—the absence of strong voltage surges, or the sudden inrush of current into a cold circuit. Equipment that is always energized escapes the shock and often lasts far longer.

Thus, one technique for extending the life of hi-fi equipment is to switch it on and off as few times as possible. This is not to suggest that equipment be left on all the time. But if you are leaving the listening room and intend to return soon, it is better to leave the switches on.

For the electronic hobbyist, a special device known as a "Surgistor" can be added to a circuit for cushioning the initial current flow. As shown in Fig. 8-3, the device is a resistance

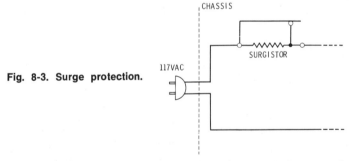

Fig. 8-3. Surge protection.

element and thermostatic arm. When the circuit is turned on, the resistance is in series with one leg of the ac line. It reduces current flow, and voltage to the circuit is limited. After a few moments, the resistance heats the thermostatic element which closes a set of contacts. The resistance is now locked out and normal line voltage is applied to the circuit. The handy hobbyist can install the device inside the hi-fi chassis in one leg of the ac line. The Surgistor, or similar accessory, is also available as an external accessory that requires no wiring installation.

75

Shifting Line Voltage

Listeners are occasionally concerned about line voltage. In some homes there are definite symptoms of fluctuating ac. When voltage runs high, lamps glow brightly and frequently burn out. An electric drill may spin faster than usual. Low voltage, on the other hand, shrinks tv pictures and the electric toaster only dries, not browns, the bread. What do these shifts do to a hi-fi system, and what can you do about them?

Consider first, the correct voltage. If you check equipment specifications, you may note that a manufacturer states something like 105-120 volts ac. The National Electrical Code (adopted by many municipalities) says line voltage should never exceed 150 volts. Because of these possible differences in house current, equipment for house-current operation will generally perform well over a sizable range. Measurements taken in homes located at several points in the U.S. have revealed that line voltage is usually somewhere in a band from about 117 to 125 volts. A daily variation of several volts is to be expected because of "load cycles," or shifts which reflect electrical consumption in an area. At midnight, for example, the line may jump up a volt or two as people retire. Greater swings occur if neighboring homes acquire heavy appliances (usually air conditioners) and the power company has made no allowance for changes in average load. (The usual remedy is to install a new transformer on the utility pole.)

A hi-fi system should suffer no ill effects with considerable changes in ac input voltage. A five percent change will probably go unnoticed. When the change is greater than 10 percent (say, from 117 volts to under 105 volts) there may be a subtle difference in sound because of amplifier power loss. Voltage above 125 may shorten the life of vacuum tubes and possibly other components (like electrolytic capacitors) in solid-state equipment. A phonograph though, will probably not vary speed under large line voltage changes since the motor is mostly sensitive to line frequency (60 Hz), not voltage.

An fm tuner might require retuning with short-term voltage shifts, though many tuners are equipped with zener diode regulators which automatically oppose frequency drift.

If your home is affected by large swings (beyond the 105-125 volt band), one resort is to complain to the power company. In such cases, the utility often attaches a recording voltmeter to your electrical service and monitors it for several days. The readings may prompt the company to improve service through adjustments at its end of the line.

There are several devices you can purchase to control line voltage in the home. Although they are mainly intended to cure shrinking tv pictures (the greatest victim of varying voltage), they can be used on a hi-fi system. Not all types are suitable. A model which automatically holds line voltage at a steady value is probably too expensive for this purpose. Another kind operates only when the line drops below a certain value and therefore offers no protection against above-normal voltage. The best choice may be the transformer listed in the electronic catalogs as "variable voltage adjuster." It has a switch and built-in voltmeter that enables you to read line voltage and step it up or down, as needed. A typical model handles an appliance or equipment rated up to 400 watts.

Line-voltage changes usually occur over a long period (several hours or more) and can be controlled by a manually adjustable transformer of this type. (Another line problem, the "switching surge" is covered in another item.)

76
Accidental Tape Erasure

Reel-type recorders have a special interlock for placing the mechanism in a recording position. This prevents accidental erasure of a tape intended for playback. Cassette machines also have an accident-prevention feature. There are two removable

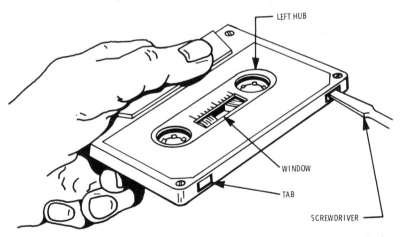

Courtesy Harmon Kardon

Fig. 8-4. Accidental tape-erasure protection.

tabs (Fig. 8-4) on the back of the cassette. To disable the record (and erase) function of the machine, commercially recorded tapes have the tabs already removed. You can do this yourself to protect valuable recordings made at home.

Insert a small screwdriver into the opening around the tab and pry it away. Don't allow the tab to fall inside the cassette. If you change your mind about the tape at a future time, and wish to record over once-valuable material, it is easy to restore the cassette. Just place adhesive tape over the opening. This actuates a lever in the machine to restore the erase function.

You can also choose the side, or track, of the cassette you wish to treat. Remove the cassette from the machine, but do not turn it over. Rotate it until you view the exposed area of tape. The tab on the left-hand side corresponds to the track that was just active on the machine. Remove the tab to prevent erasure, or cover it to restore recording.

77

Sticky Cassette

When a cassette is stored on a shelf for long periods, the layers of tape may stick to each other. Such tapes play with erratic tension and cause annoying speed changes heard as wow and flutter. The treatment is to run such tapes through the transport at fast forward speed, then again at fast rewind. This eliminates the sticky sections and may be done for either raw or prerecorded tapes, especially if they have been on a dealer's shelf for several months.

A similar problem is avoided by not rewinding a cassette immediately after recording or playing. It should be stored in this condition because tension between tape turns is most uniform just after running through the machine at slower playing speeds. Rewinding is done immediately before the next play or record session.

SECTION 9

Accessories

78
Simple Crossover

A crossover network is a filter which splits the audio output of an amplifier into frequency bands. In the usual arrangement, low tones are applied to the woofer, the high tones to a tweeter. Crossovers contain large coil and capacitor combinations or active electronic circuits. Here is one of the simplest crossovers you can use, but one that does a remarkably good job. It is handy when you wish to add a tweeter to an existing loudspeaker system to improve high-frequency response.

As illustrated in the diagram (Fig. 9-1), the crossover is actually a capacitor in series with one side of the amplifier line. The capacitor is a high impedance, or obstacle, to low tones, while permitting highs to go through nearly unaffected. The tweeter is also protected by the capacitor against overdrive

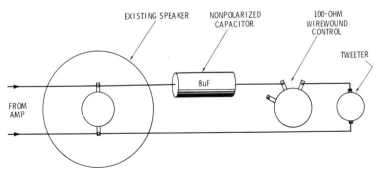

Fig. 9-1. Simple crossover network.

103

from the low end of the audio spectrum, where most wattage is concentrated.

The capacitor is a nonpolarized type of approximately 8 microfarads. (A standard electrolytic capacitor should be avoided since it is designed for dc operation. The speaker signal is ac.) Using an 8-ohm tweeter, the approximate crossover point will be at 3000 Hz. The tweeter may be a speaker designed for the purpose, or you can mount a small speaker in its own box to achieve nearly the same effect. A speaker with a diameter of about four or six inches should work.

Also shown in the diagram is a brilliance control. It is an optional item, but it gives an added measure of control over the tweeter. After the system is in operation, adjust it for the most pleasing sound. If you do not install a brilliance control, use the treble control on the amplifier to adjust the highs.

79

Test Record

Here is a handy item for judging the overall condition of a hi-fi system. It will not replace a technician's instruments or provide exact specifications, but it gives an approximate idea of how equipment is functioning. In the example shown in Fig. 9-2, the disc is divided into stereo and mono sides. On the mono side, for example, two bands give an indication of stylus wear. Play each band in turn, and listen for equal loudness and clarity. The test also reveals a dirty cartridge or particles jammed into its crevices.

In another band, a voice announces several tones from 30 to 15,000 Hz. If an output meter is connected across speaker leads, a steady reading indicates correct frequency response

Fig. 9-2. Test record.

and RIAA equalization. Two tones played simultaneously in another band reveal intermodulation distortion, heard as fuzziness or a background noise. A rumble test is done with a very low tone, followed by silent grooves. Tone-arm resonance is accomplished by a tone that sweeps from 50 Hz down to 10 Hz. A good tone arm and pickup tracks the tones without jumping grooves.

On the stereo side of the record, you can get a good idea of stereo separation. In switching tones between channels, the disc lets you monitor interaction or distortion between them. It exposes troubles like a misaligned cartridge. A clicking sound of a metronome recorded equally on both channels is also provided for adjusting amplifier controls for exact volume balance. It is done while you sit in your favorite listening chair, while someone adjusts the controls. The clicking noise appears to emanate at a center point between the speakers when everything is balanced.

80

Hiss Reduction

A severe limitation in the quality of cassette and other slow-speed tapes is hiss. It is most annoying on soft passages and solo instruments. The reason is that a tape recorder has a fixed, internal noise level which competes with a low-level signal during playback. A vast improvement is possible by processing the signal through special noise-reduction circuits. Most well-known of these systems is the "Dolby," widely used in the professional studio for recording original performances. It divides the total audio signal into four frequency bands and senses their level. Any band which falls below a given strength is selectively boosted. When it goes through the recording phase, the processed signal is far stronger than that of inherent noise. During playback, the process is reversed to bring the "stretched" signal back to its original proportion. This not only gives weak signals greater resistance to noise, but less hiss, rumble, and print-through are transferred to commercial recordings copied from master tapes.

Devices at much lower cost also provide noise reduction for the home recordist. Such circuits may be installed in the original recorder, or used as an accessory (Fig. 9-3) inserted between the program source (a tuner, for example) and the recorder through standard plug-in cables. They derive lower cost by eliminating three of the four channels, or frequency bands, found in studio models. The circuit deals solely with the high-frequency end of the audio spectrum, starting at about 1700 Hz where the poorest signal-to-noise ratio exists. Any signal in this region which falls below a given strength is boosted, then reduced during playback to lower the hiss levels.

Courtesy Advent
Fig. 9-3. Tape-noise reduction center.

81

Audio Mixer

This is about the simplest mixer you can build (Fig. 9-4). It blends two audio signals in the desired quantity and combines them into a single output. You can mix signals from a tuner, tape recorder, or other audio source and add some professional touches to your home recording.

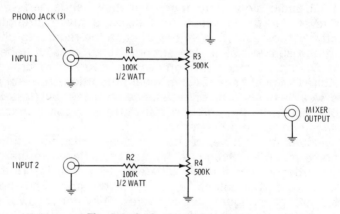

Fig. 9-4. A simple audio mixer.

Since the circuit is so simple, with no electronic amplification, it has limitations. Signals to be mixed—the output of a tuner, preamplifier, tape preamplifier, or voltage picked from across loudspeaker leads—should be introduced at about the same voltage levels. All these signals measure about a volt or so and are compatible. Two microphone outputs can be mixed together since their levels are also similar (measured in a few thousandths of a volt). Avoid direct pickup of a magnetic phono cartridge since it requires equalization networks. You might, however, use the output of a preamp that contains the phono signal.

The mixer is built in a small aluminum case for shielding against hum pickup. The two inputs and single output are phono-type jacks. The 500,000-ohm potentiometers are carbon controls with an audio taper. Keep all leads short. If you discover that the pots rotate in the wrong direction (clockwise motion should increase volume), reverse the leads going to the outer lugs of the control.

82

A-M Radio Adapter

Fm is considered the medium for hi-fi broadcasting, especially since it transmits stereo, and many tuners omit a-m coverage entirely. To fully exploit a hi-fi system, you might wish to add a-m reception. Contrary to what many people believe, a-m broadcast stations are not severely limited in frequency response. They transmit more than is heard on a table model radio. A-m extends the number and type of programs you can hear. Fortunately, it is easy to add it to an fm only system.

One fault of a-m occurs during nighttime reception. Due to the ability of a-m signals to "skip" great distances, much interference is apt to be heard unless you are tuned to strong stations. Further, when listening to a-m through a hi-fi system, you may experience the phenomenon known as the "10-kHz whistle." It occurs when a distant station arrives and mixes with a second station to produce a difference tone (or heterodyne). Since the stations are 10 kHz apart, an audible high-pitched tone is heard. The problem won't happen during the day since distant stations cannot be heard. Any two local stations are always spaced at least 20 kHz apart—too far to produce the same effect.

One of the simplest approaches to adding a-m is an old-fashioned crystal radio. A typical circuit is shown in Fig. 9-5, but they are widely sold in electronic supply and department stores (in the toy or hobby section.) The circuit has a broadcast antenna coil, a 365-pF variable- tuning capacitor, a diode de-

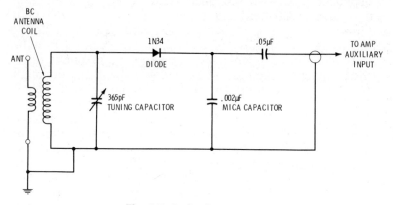

Fig. 9-5. A simple a-m tuner.

ector (1N34) and a small filter capacitor of about .002 µF. You can assemble the circuit from a low-cost kit, or purchase these parts and follow the diagram.

Almost invariably, the new builder of a crystal set claims "It doesn't work!" And almost invariably the reason will be a poor antenna. At least 30 feet of wire must serve as an antenna for local stations; more length if you want greater range. A good electrical ground is also critical. The circuit has absolutely no amplification and depends on air signals for power. The a-m signal from the circuit is fed into an auxiliary or tuner jack on the hi-fi amplifier.

The sound of a crystal radio is often quite good (it has so few parts to affect the signal!). A drawback is selectivity. If strong stations are close to each other on the dial, they may be heard simultaneously.

Another approach to adding a-m is with a pocket-size portable receiver. These are very inexpensively available. The sound from such receivers is usually atrocious, but their selectivity is vastly better than that of a crystal circuit. Further, it is possible to detour the section of the receiver that wrecks audio quality. First thing to try is a connection from the earphone jack into an auxiliary input on the hi-fi amplifier. It bypasses the tiny internal speaker for some improvement. To get even better performance, you can bypass the audio amplifier in the radio and pick up the signal at the detector. You shouldn't need a schematic since the signal is tapped from the volume control. As shown in Fig.

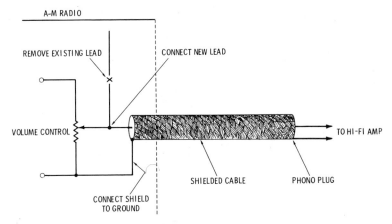

Fig. 9-6. Connecting an a-m radio to a hi-fi.

9-6, the lead to the center terminal of the volume control is removed and a shielded lead connected in its place. The shield is connected to the radio ground, which is one of the outer terminals of the volume control. If you can't identify the ground terminal, touch the shield to either outer lugs to see which produces best results through the hi-fi system.

83
Cooling Fan

The changeover from tubes to transistors in hi-fi equipment greatly reduced the amount of generated heat. It was a welcome improvement since high temperature is a notorious cause of short component life. Solid-state designs, nevertheless, produce heat, as evidenced by large finned radiators on output transistors. Several other factors in solid-state offset the lower heat advantage—cabinets are smaller, with less radiating surfaces, printed-circuit boards do not conduct heat as well as metal chassis, and circuits are built with higher component densities. If you find your hi-fi cabinet uncomfortably warm to the touch, a cooling fan may prove a worthwhile accessory. Not only will circuits run cooler, but a fan lets you mount equipment in tighter spaces where natural air flow is restricted.

Choose a fan model made for hi-fi cooling. Its blades are designed to produce little wind noise and least disturbance to the music. Select a mounting location that enables the fan to provide the most effective circulation. Best place to plug the fan into ac power is a switched outlet on the hi-fi amplifier. It automatically operates the fan whenever the hi-fi is turned on.

84
Recording Adapters

It is a poor practice to record from a radio, tv, phono or other program source by placing a microphone near a speaker. Not only does it reflect loudspeaker deficiencies, but room noises interfere with recording. Acoustics and reverberation also distort the sound. It is far better to pick up the signal through direct electrical connection into the circuit. The simplest pickup is at the loudspeaker terminals, as shown in Fig. 9-7A. Run a pair of wires to the speaker and solder them to the terminals, or use alligator clips to make a temporary connection for recording. Twist the wires and attach a phono plug, as shown, for inserting into the tape recorder. If the line is more than about 10 feet, it might require a shielded type to prevent hum.

Another adapter, seen in Fig. 9-7B must use shielded cable regardless of length. This installation is more difficult because

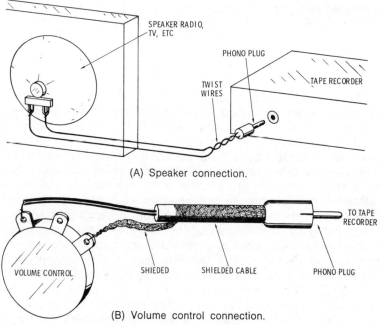

(A) Speaker connection.

(B) Volume control connection.

Fig. 9-7. Recording directly.

you must gain access to lugs on the volume control. The two outer lugs of the control receive the shielded and center leads from the cable to the recorder. If you experience hum, reverse the leads to the outer lugs of the control. This tapoff is better than the speaker connection since it eliminates all audio amplification in the radio or tv. (The signal is received from the detector stage.)

In the first adapter, using speaker connections, you will still hear normal audio in the radio or tv speaker. Also, the volume control on the radio or tv controls recording level. In the volume control hookup, full signal is available even when the radio or tv volume is turned to zero. Level adjustments are done exclusively at the recorder.

In both hookups, use the high level (auxiliary, tuner, etc.) input of the tape recorder. One precaution must be observed, if the radio, tv, or other source is rated for ac-dc operation, do not use either of these adapters because of possible shock hazard. They are safe, however, with transformer-operated devices if all bare wires are insulated against short circuits.

85

Station Guide

Daily newspapers give program and frequency information for local fm stations. The trouble is, sensitive fm receivers and

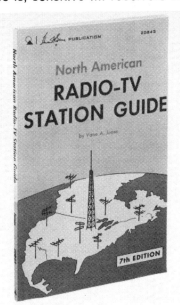

Fig. 9-8. An example of a station guide.

elaborate outdoor antennas may extend receiving range beyond 100 miles. If you wish to listen regularly beyond local limits, a publication like the *North American Radio-TV Station Guide* (Fig. 9-8) is handy to keep near the receiver. It lists all fm stations by geographic location and frequency. The information includes antenna height, transmitting power, whether the station broadcasts in stereo, and some indication of programming. Once you have located stations you want to hear, you can write to them in care of their call letters, town, and state. Many stations publish periodical guides with complete program details.

SECTION 10

Troubleshooting Tips

86
Intermittent Kits

Since a considerable amount of hi-fi equipment is built from kits, consider the major cause of intermittent operation in these circuits. The service department of a leading manufacturer reports that most kits returned to the factory for repair are troubled by one ailment—cold solder joints. Many of these kits are restored to perfect operation by the simple expedient of resoldering every joint on the chassis. So, try this cure before submitting equipment for repair.

In nearly all cases, the powdery, crumbling cold-solder connection is caused by failure to heat the complete joint—both wire and terminal—to a temperature where either can melt solder. Be sure to touch the iron tip to both simultaneously during the heating period. Also, let the joint, not the iron tip, cause solder to melt and run. Follow kit instructions faithfully and you will get a trouble-free connection every time.

87
Tube Testing

In tube equipment, the passage of time produces a multitude of effects on performance. Yet, it is impossible to recommend an exact timetable for changing, or even checking, tubes. Tubes often deliver excellent performance well beyond their normal life expectancy. One study has shown that catastrophic failure (which halts equipment operation) mainly occurs during the first one hundred hours of operation. Such failures are very low after 500 hours. Nevertheless, there are certain circuits that are particularly vulnerable to tube aging and deserve special attention.

High current tubes in the output of an audio amplifier, for example, suffer a loss of emission in time. The drop can reduce amplifier power and cause distortion while driving low-efficiency loudspeakers. Another heavy current tube is the rectifier in a power supply. If this filament emits fewer electrons there could be a high-voltage loss along the B-plus leg of the supply. It reduces overall wattage and possibly causes a shift in delicate bias voltages on tube grids.

Another tube to watch is the bias oscillator of a tape recorder. If it malfunctions, there could be an increase in recording distortion. The ultrasonic signal generated by the tube may emerge distorted or too feeble to adequately perform the bias function.

A critical stage in audio equipment is the preamplifier. It is the first tube in a magnetic phono and tape recording amplifier. Because the tube operates at very high gain, defects are given a significant boost. If you hear random static or hiss that disappears when you remove the preamp tube, try replacing it. (The problem, incidentally, may not show up on a tube tester.) Similarly, the first stage in an fm tuner—the rf amplifier—can also add noise in time.

On tube checking: the drugstore tester is best for determining catastrophic failure, like a burned-out filament. It also gives an indication of a short-circuit between tube elements. The readings are only approximate for emission. The ability of a tube to amplify within its rating would have to be indicated on the superior mutual conductance checker, usually found in service shops. If you wish to purchase spare tubes, then substitution in the equipment itself is a fine troubleshooting technique. (And having those spares on hand lets you make a fast repair when the store is closed.)

How often should you remove tubes for testing? Some authorities recommend an annual checkup. This might prove too frequent, and a two-year schedule sounds more practical. If you do find a tube which tests low in emission, do not expect dramatic results when you replace it with a new one. Differences, especially in audio power, are difficult to perceive with the ear. Low emission, however, might be a sign that the end of a tube's useful life is near.

88
Cartridge Terminals

Connections to a phono cartridge form a weak link in the signal chain. They are usually fitted with friction, or slide-on clips which occasionally fail to maintain good contact. Common symptoms of trouble are hum and no signal. In the former, the clip which grounds or shields the cartridge fails to make good contact or has slipped off. Loss of signal occurs when one of the "hot" clips fails to make proper contact. Lack of tension in the clip may also introduce circuit resistance which alters the audio signal.

Fig. 10-1. Check terminal connections on cartridge.

Remove the cartridge and visually check the terminal connections (Fig. 10-1). Do not solder clips to the cartridge pins; it could melt the plastic material. Shine the pins with a cloth to remove dirt and to assure a good electrical contact. If contact friction is poor, pinch the clip slightly with a small pliers before sliding it onto the pin.

89
Ground Loops

Various chassis and cabinets of a hi-fi system are usually grounded to each other through shielded signal cables. In some instances, connecting more than one ground between components can introduce a ground loop, which is heard as a 60-Hz hum in the speakers.

Although grounded chassis are considered to be electrically zero, a multiple ground can cause slight unbalance, or difference in potential, between chassis. It is the result of circulating hum currents in the chassis that encounter a tiny amount of electrical resistance between ground points. A voltage drop occurs across the resistance, and this drop is amplified as a normal signal in the system.

Manufacturers of hi-fi equipment avoid such loops inside a chassis by careful choice of grounding points. Generally, the technique is to ground all circuits to just one point. This prevents multiple paths from creating undersired loops. A similar approach may be used with components in a hi-fi system. If you have grounded several chassis together with an external wire, remove it to see if it reduces hum level. Normal signal cables may provide all the required grounding. Also, check if cabinets are touching each other, and possibly introducing additional grounds into the system. If there is a ground wire running to a cold-water pipe, disconnect it, or change its location, to note any effect on hum.

90

Hidden Cable Breaks

Two common troubles in hi-fi systems—loss of signal and hum—often prove to be mechanical, not electronic, problems. They are traced to defective cables which run between various components or chassis. Try these simple checks, each done with the system set up for normal listening.

The most vulnerable spot is where a connector attaches to a cable—usually a phono plug molded on, or soldered to, a shielded phono cable. Grasp the cable near the plug and twist, bend, wiggle, pull, and othewise stress the cable. If there is a break in the shield at this point, chances are you will restore contact temporarily and the hum should disappear. The same treatment is done in the no-signal symptom. Disturb the cable as just described, and you may hear the program. If you cannot locate the trouble near a plug, bend and wiggle the cable along its entire length (Fig. 10-2).

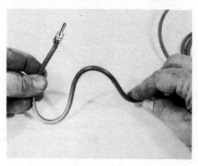

Fig. 10-2. Checking for broken wires.

A practice that causes internal cable breaks is removing a phono plug from its socket by pulling on the cable. Although manufacturers do not put handles on their products, always grasp the plug, not the cable.

91
Scratchy Controls

After several years of operation, the volume, tone, balance, and other controls in a hi-fi system may produce an audible scratching noise in the speaker as they are turned. It is caused by entry of air-borne grease particles into the control, or loss of spring tension in a contact arm which slides against a carbon element. When you operate the control, it fails to maintain perfect contact and a scratchy noise is heard. As the problem grows worse, poor contact introduces additional resistance which can cause audible hum. Another sign is a temporary loss of signal as the arm fails to make contact. Tapping the control may restore normal operation for a while.

These symptoms mean that a control may soon require replacement. It is not possible to tighten contact tension. Additional life, however, is often achieved by spraying the control with an aerosol contact cleaner, like the one pictured in Fig. 10-3. The ac power is turned off and the spray is directed into an opening on the case of the control. As the spray enters, keep the control in motion by turning the knob back and forth quickly. It assures good cleaning action and even distribution of the cleaner. Although our photo shows the control removed from the circuit, you should be able to spray it while it is mounted in place. For inaccessible areas, use the extension tube supplied with many spray cans. Other items with moving contacts —slide switches, rotary switches, and toggle switches—can also be treated. Switch them back and forth as you spray.

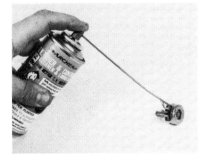

Fig. 10-3. Cleaning a noisy control.

92

Free Test Set

Servicing a television receiver is greatly simplified since each set has a built-in oscilloscope—the picture tube. Many troubles produce telltale symptoms on the screen. Stereo amplifiers have a built-in tester of even greater value—the other channel. Much troubleshooting can be done by simply comparing an ailing channel against the healthy one. Not only are channels electrically the same in virtually all cases, but they have identical physical layouts. It makes test points easy to locate and compare.

What if *both* channels do not work? This isn't serious since it localizes the trouble to the power supply, a circuit common to both sides of the stereo amplifier.

Let us say one channel is inoperative. If you have nothing but a voltmeter for test equipment, operate the amplifier, but apply no signal to the input. Now compare dc voltages at identical points in each channel. Readings should be within approximately 20 percent of each other; e.g., with a 10-volt indication at some point, anything from 8 to 12 volts at the companion point on the other channel will probably fall within acceptable tolerance.

If you have more sophisticated equipment—a signal generator and oscilloscope—you can introduce a signal simultaneously to both channels. The scope is used to check signal quality and strength at comparable points in the circuit. Since tubes and transistors are prime suspects, you might interchange them between good and bad channels to verify their condition.

SECTION 11

Interference

93
Electrical Ground

Interference to a hi-fi system springs from many sources, as shown by several items in this section. A good starting point in attacking interference, however, is to provide a proper electrical ground. Components of a hi-fi system may resist interference through shielding and bypassing, but fail when interconnected. The reason is usually that voltages which create interference cannot easily drain to ground.

In other cases, long cables between cabinets act as antennas which pick up radio signals or air-borne disturbances radiated from electrical machinery. Proper grounding lowers these possibilities.

A good electrical ground is sometimes difficult to achieve. It may, in fact, worsen a hum condition. So consider the following suggestions as a starting point, then try some experimentation on your own. It is not too difficult to discover the combination that renders interference least objectionable. Later, you can deal with the problems described under specific headings.

Before attempting any grounding read the "ground loop" entry in another section. It tells how to eliminate hum which arises from too many grounds. Solve this problem before trying to install the electrical ground being described here. Once the system is quiet, you will tie it into the same electrical ground which serves the ac house wiring. It will be attached to a point in the hi-fi system which produces least hum. That point may be already labelled as a ground by the hi-fi manufacturer. Otherwise, a chassis or cabinet screw may be loosened to receive an external ground wire. The wire can be a bare lead of No. 18 or heavier, solid copper wire.

It is assumed your hi-fi system is not of the ac-dc type (a rarity in high-quality equipment) and the chassis isn't "hot." Check the manufacturer's literature on this. Also avoid using any screw which holds a finned heat sink to a chassis. It might bear some value of transistor voltage that could be short-circuited by a connection to ground.

A good electrical ground is often found at a cold-water pipe. You can run a ground wire to the pipe and use a special clamp (made for the purpose) to make the connection. Sometimes pipe joints are coated with dope and a continuous electrical ground is not available. Also, a water pipe might be too far removed from the hi-fi system. Another choice for a ground would be the screw which holds the cover plate on an ac wall outlet. In almost all cases, this screw contacts a junction box inside the wall which connects to electrical ground of the building. Since hi-fi equipment plugs already run to this point, adding the ground wire is usually simple to do with a spade lug slid under the cover-plate screw.

If none of these alternatives is practical and grounding is clearly needed, obtain a copper grounding rod and drive it into earth. It has a terminal for accepting the ground wire.

When interference is mainly to fm radio, read the section devoted to fm reception. Most fm interference problems disappear if you take steps to improve incoming signal strength. Interference is almost always a-m (amplitude modulated) and should be rejected by the limiter and detector circuits in the fm tuner (which respond almost exclusively to fm, not a-m.) If fm signals are weak, however, noise-limiting action is poor and interference more easily disturbs reception.

CB-Ham Voices

When you hear voices on your amplifier but aren't playing tuner, tape, or records, it is probably interference from a nearby CB or ham radio transmitter. Legally speaking, it is not often the CB or ham operator who is at fault since these stations may be operating well within federal regulations. It is the hi-fi rig that is insufficiently protected. Two common entry points for these signals are the ac line cord and the speaker leads.

Many cases are cured by short-circuiting the path for the radio-frequency signal. As shown in Fig. 11-1A, a bypass capacitor is soldered from each leg of the incoming ac line, to ground, near where it enters the chassis.

The radio signal path on loudspeaker lines is shorted by placing a capacitor from the "hot" speaker lead to the chassis (Fig. 11-1B). The capacitors shown for ac and speaker-line bypassing will not affect the normal operation of the circuit. If these steps do not completely kill the interference, proceed to the next step.

Your circuits may be suffering a bad case of "grid-circuit rectification." That's an old term used to describe what happens when a strong radio signal reaches the grid terminal of an audio tube. The signal sees a diode formed by tube grid and cathode, and it starts a detection process, much like that of an a-m radio. Intelligence is extracted from the radio signal and audio is

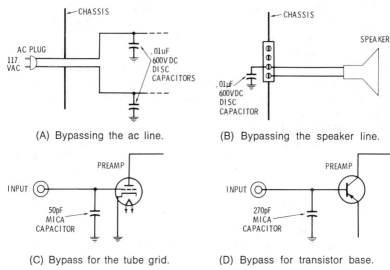

Fig. 11-1. Eliminating CB interference.

amplified and heard as interference. The usual cure is a small capacitor added from grid to ground to short-circuit the radio energy. The hookup is shown in Fig. 11-1C for a tube grid. A 50-pF mica capacitor should do the job without affecting the regular function of the stage. The bypass capacitor is usually placed in the first tube grid, often the phono preamp stage.

If your equipment is solid-state, and it suffers from interfering voices, the problem must now be called "base circuit rectification," since transistors do not have grids. The problem is still the same, or possibly worse, since the input of a transistor is already a diode circuit. Radio signals are readily detected into audio at this point. The bypassing technique, shown in Fig. 11-1D, consists of a 270-pF mica capacitor installed between the base of the first transistor (the preamp) and ground. A larger capacitor than used in the tube circuit is selected because of different input impedances between tube and transistor circuits.

95

Transmitter Filter

Harmonic interference is a disturbance occasionally suffered by fm receivers. The intruding audio, which may be heard at one point on the dial, is often caused by a powerful transmitter operating nearby. (Transmitters produce other types of interference, described elsewhere in this section.) The operating frequency of the transmitter may be lower than that of the fm band, but harmonics (or multiples) may fall within the 88 to 108-MHz band. For example, the second harmonic of a 6-meter amateur radio rig falls somewhere between 100 and 108 MHz. Police and fire radio services occupy a 30-50 MHz band, which might also prove troublesome if the transmitter is located very close to your fm receiver.

Curing harmonic interference can not be done at the fm tuner. Harmonics are too close in frequency to fm broadcast stations to be filtered. The problem must be treated at the source. If it is a public-safety service (like police) there is probably little you can do. But if interference is known to be from a local ham operator, you might persuade him to install a *low-pass* filter on his transmitter. The device (Fig. 11-2) in its most popular version, prevents signals higher than 52 MHz from being transmitted.

In most instances, the ham is not compelled by law to install such a filter. (The Federal Communications Commission has

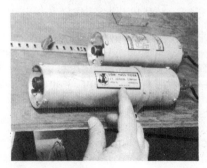

Fig. 11-2. Low-pass transmitter filter.

discovered that most faults lie in the receiver, not the transmitter.) But it is probably in the ham's best interests to install the filter. If he is disturbing your fm reception, he is also probably tearing up tv pictures all over the neighborhood.

96

Arcing Motors

Count the number of motors in the home today and you will often find a dozen or more. Heavier motors, which have carbon brushes, are common sources of interference to radio reception. Typical items include mixers, vacuums, and power tools. It is easy to identify the source—the interference stops when the appliance is turned off. Although the disturbance most often affects receivers, it can be strong enough to directly enter an audio amplifier and be heard in the loudspeaker.

Interference from large appliances is caused by sparking at brushes. Since wear and dirt aggravate the condition, you can improve matters by cleaning the offending source. Remove the appliance plug from its outlet, and open the case to expose the carbon brushes and commutator (a circle of copper bars or segments). Blow out dust or other particles in the case. The commutator will probably be darkened with carbon, so use very fine sandpaper to return the shine to the copper bars. Do not get metal between the copper segments since these spaces must remain an electrical insulator. For this reason, avoid emery paper for the clean up since it could short-circuit the bars with conducting particles.

If the brushes are badly worn, or their springs suffering from loss of tension, replace these items. Use alcohol on a cloth to shine up all parts, and reassemble the motor in its case. Not only will you have cured interference, but the appliance will probably run better, too! If some static persists, you can add bypasses, described elsewhere in this section.

97

Appliance Bypasses

These capacitors are small ceramic disc types which shunt radio frequencies to ground before they can enter hi-fi equipment to cause hash, buzzing, or other interference. The simplest bypass capacitor is shown in Fig. 11-3A. It connects across the ac line. (All the bypasses shown are .01-μF units rated at a minimum of 600 volts.) This, incidentally, is the low-cost interference filter widely sold in stores. It plugs into an ac outlet, and the appliance cord plugs into the filter. You can do a better job, however, by mounting the bypass capacitor inside the appliance if there is available space. This location eliminates the appliance line cord as a radiating antenna.

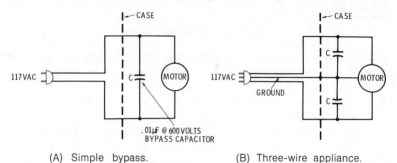

(A) Simple bypass. (B) Three-wire appliance.

Fig. 11-3. Appliance bypass capacitors.

Two capacitors are used in the 3-wire appliance (which will ultimately replace the 2-wire types). Be careful to tape any exposed wires inside the appliance case and check that no rubbing can cut any wires and cause a short circuit.

If heavy office equipment (cash register, adding machine, etc.) is operating nearby, a good interference suppressor is the choke-type line filter. This device, sold in radio supply stores, requires a good electrical ground, and must have sufficient capacity in amperes to carry the total appliance current.

Curing interference is most successful when it is treated at the source. In many cases you will not be able (or willing) to bypass a neighbor's appliance. Finally, check your hi-fi equipment for bypassing at the ac line, just after it enters the chassis. Add a bypass capacitor across the line, if none was installed by the manufacturer.

98

Aircraft Interference

A complaint heard from some fm listeners: "I hear airplanes through my hi-fi!" This brand of interference may occur even when the listener is far from an airport. At high altitude, the transmitting distance of aircraft is more than 200 miles. Also, an airplane flying over the general listening area can be less than six miles away, in the vertical direction, from the fm an-

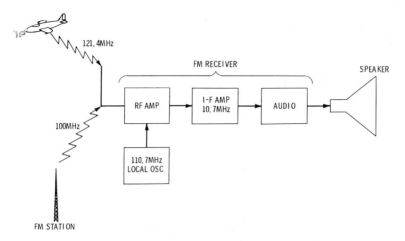

(A) Formation of image frequency.

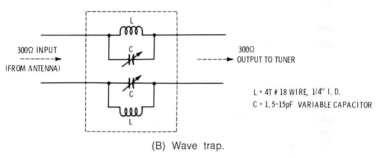

(B) Wave trap.

Fig. 11-4. Eliminating aircraft interference.

tenna. The reason behind most aircraft interference on fm is the "image." It is a false signal created in the fm tuner because of the frequency band assigned to air traffic.

How the image is formed is shown in Fig. 11-4A. As you may know, an fm tuner converts all incoming signals to 10.7 MHz, an intermediate frequency that permits good selectivity and amplification. The step-down in frequency is aided by the local oscillator which mixes with the incoming signal. In our example, the desired fm station is on 100 MHz; it mixes with the 110.7 MHz oscillator signal and the difference—10.7 MHz—is amplified in the i-f stages.

Note that the airplane is transmitting on 121.4 MHz, a frequency in the middle of the aircraft communications band. The problem begins when the signal enters the fm receiver and, due to a mathematical coincidence, also produces a 10.7-MHz signal in the receiver i-f amplifier. (The aircraft signal is a-m but it is often heard with excellent clarity on fm sets, anyway.) There are several cures for the image.

Anything that improves the selectivity of the receiver "front end" aids in image rejection. For example, early solid-state tuners were especially poor in selectivity. Later designs were sharpened with a double-tuned rf amplifier. (The way to see this is to count the number of "gangs" on the variable tuning capacitor. Better tuners often have four, not three, sections.) Another device to reduce images is the FET (field-effect transistor). If you live in an area with high-density air traffic, consider these design improvements when purchasing a new fm tuner.

Another way of narrowing front-end selectivity—and thus reducing images—is to improve the antenna system. Listening on a single wire, a line-cord antenna or rabbit ears may supply adequate signal strength on fm stations, but such antennas may pick up aircraft signals just as well. A yagi antenna or folded dipole (described elsewhere) cut for the fm band adds an additional tuning circuit that sharpens the receiver.

In bad cases, a wave trap may sufficiently attenuate the offending signal. The circuit shown in Fig. 11-4B is one that may be tried by an experimenter with some experience in vhf frequencies. (It is too critical for the beginner to construct.) The trap consists of a series-tuned circuit in each leg of the 300-ohm lead-in from the fm antenna. It may be assembled on a small piece of phenolic circuit board and mounted as close as possible to the antenna terminals of the fm tuner. (Construction values are given in the illustration.) To prevent the coils from interacting with each other, their ends should be mounted at right angles.

To adjust the wave trap, turn on the fm receiver and set the two tuning capacitors to about half capacity. A signal generator is used to simulate the image frequency on about 120 MHz. This should be tuned and heard in the receiver so use a tone-modulated signal. A nonmetallic instrument is used to pry apart or compress coil turns until the image signal is noticeably affected. Spacing the coil turns gets the trap into an approximate 120-MHz range, and you continue fine tuning with the variable capacitors. Go back and forth to check for best trap alignment, while leaving normal fm reception unaffected.

99
High-Pass Filter

If a transmitter is located very close to the fm receiver, it could cause a condition called fundamental overload. Powerful signals, though far removed from fm frequencies, enter the receiver and are heard throughout the band. In some cases, the signal detours around the front end and enters the intermediate or audio stages. Such signals may be from ham, CB, police, fire, radio-paging, diathermy, X-ray machines, neon signs, or other sources of radio-frequency energy. Much of this interference can be minimized with a high-pass filter.

The same high-pass filter sold for reducing television interference can also be used for fm reception. The most popular version is the 300-ohm high-pass type which connects directly in the lead-in between tuner and antenna (Fig. 11-5). Electrically, it blocks all frequencies from 0-52 MHz. Since tv station frequencies begin just above 52 MHz, and fm commences at 88 MHz, the filter leaves these signals alone. Interfering signals below 54 MHz, however, are sharply attenuated.

Note that a high-pass filter cannot cut interference that falls within the 88 to 108-MHz band; nor will it cope with "images" caused by even higher frequency signals (described in another item).

Fig. 11-5. A 300-ohm high-pass filter.

In mild cases of interference, the filter is attached at the antenna terminals of the fm tuner. Where transmitter signals are extremely strong, the filter should be installed within the fm cabinet, if room is available. The filter is symmetrical, that is, it operates when connected either way. It is worth trying both directions, though, to note any possible improvement. If your fm antenna system uses coaxial cable use a 75-ohm high-pass filter.